"Rob Ryan has a passion for business and a love for entrepreneurs. His gut entrepreneurial instincts are right on and he communicates them in a persuasive, in-your-face manner. Our interactions with Rob have always resulted in higher expectations and a practical plan to achieve them."

— Greg Gianforte, CEO, RightNow Tech

"The founder of Ascend Communications, having sold the company, now spends his time counseling other entrepreneurs. The result is this chatty, concise guide aimed at people so infatuated with their brilliant ideas that they risk losing sight of market practicalities. Ryan covers the entire start-up process, from focusing your business plan on what paying customers actually want, to pitching only to carefully targeted venture capitalists, to quickly adding seasoned managers who will keep you ahead of the competition. Despite the folksy language, the book is full of smart suggestions."

—John T. Landry, *Harvard Business Review*

"Behind the scenes of many hot MIT and Stanford startups, Rob Ryan is getting his hands dirty for the firms he mentors. At last, he's distilled his lessons into a remarkably readable and immediately useful set of battle plans. *Entrepreneur America* is advice from a High General of entrepreneurship: Rob is the architect of Ascend Communications, the world-class network hardware firm that grew to over $20 billion in value. This book is more than the *Art of War* for the office: it combines strategy with proven examples, such as actual IPO presentation slides and model board meeting notes, as well as step-by-step campaigns for meeting customers, building products, and raising cash. It's a super tool for folks who haven't had the opportunity to attend the Montana boot camp. It's a potent tactical guidebook, and provides the framework for building a sustainable enterprise. Do yourself a favor: Read this book before your competitors do."

— Gregg Favalora, founder / CTO of Actuality Systems, Inc., manufacturer of True 3-D Display systems

"Rob isn't just a savvy businessman; he's an entrepreneur. The alpha-entrepreneur. Rob's mentorship has helped guide [PatientKeeper] from a prefinanced venture to a formidable company that has reshaped the e-health landscape."

— Stephen S. Hau, founder, PatientKeeper

"When I first met Rob he told me that he is right in 98 percent of cases. Over a year of close interaction with him, I've had many opportunities to test that statement, and now I know that it was precisely accurate. Working with Rob allowed us to gain the executive experience and maturity that our entrepreneurial management team needed so much."

— Andrew Feinberg, CEO, NetCracker

Smartups

SMARTUPS

LESSONS FROM

ROB RYAN'S

ENTREPRENEUR AMERICA

BOOT CAMP

FOR START-UPS

ROB RYAN

With a new preface

Cornell University Press
Ithaca and London

First published 2001 as *Entrepreneur America: Lessons from Inside Rob Ryan's High-Tech Start-up Boot Camp* by HarperBusiness

Published 2002 by Cornell Paperbacks

Printed in the United States of America

Library of Congress Cataloging-in-Publication Data
Ryan, Rob, 1948–
    Smartups : lessons from Rob Ryan's Entrepreneur America boot camp for start-ups : with a new preface / Rob Ryan.
      p. cm.
    Paperback ed. of: Entrepreneur America : lessons from inside Rob Ryan's high-tech start-up boot camp. New York : HarperBusiness, 2001.
    Includes index.
    ISBN 0-8014-8831-1 (alk. paper : pbk.)
    1. High technology industries—United States—Management. 2. New business enterprises—United States—Management. 3. Entrepreneurship—United States. I. Ryan, Rob, 1948–. Entrepreneur America II. Title.

HD62.37 .R93 2002
620'.0068—dc21   2002073828

Cornell University Press strives to use environmentally responsible suppliers and materials to the fullest extent possible in the publishing of its books. Such materials include vegetable-based, low-VOC inks and acid-free papers that are recycled, totally chlorine-free, or partly composed of nonwood fibers. For further information, visit our website at www.cornellpress.cornell.edu.

Paperback printing   10 9 8 7 6 5 4 3 2 1

To Terry, my wife of thirty-three years
You are the wind under my wings

Contents

Foreword

In 1989 Rob Ryan founded Ascend Communications with three engineers. Rob served as CEO of Ascend, taking it public Friday the thirteenth, May 1994, at $13.00 per share. Under Rob's leadership, Ascend had become the leading manufacturer of Point of Presence boxes (POPS) for Internet providers. Rob describes Ascend's business as "selling the picks and shovels for the Internet gold rush."

In 1995, the last year Rob served as CEO of Ascend, the stock was acknowledged as the best performer of the year on all of Wall Street, returning a whopping 721%.[1] If you had invested in Ascend two months after the IPO, you would have gained 3,223% one-and-three-quarters years later.[2]

Rob Ryan's current focus is his Entrepreneur America Mentors LLC facility at Roaring Lion Ranch in Hamilton, Montana. Entrepreneur America is an incubator for high technology start-ups. In this Rocky Mountain setting, Rob mentors promising entrepreneurs, assisting with focusing product ideas, writing business plans, sharpening presentations, and raising venture capital.

In its first four years, eighteen companies have been founded and mentored by Entrepreneur America, three of which have already achieved billion dollar valuations. With

[1]*Business Week*, December 25, 1995; pp. 126–127
[2]*Investors Business Daily*, August 30, 1996; pp. A3

that record, it is no wonder that the chance to work with Rob is sought by hundreds of young technical entrepreneurs.

What is not well known about Rob Ryan is that, early in the history of Ascend Communications, his venture capital backers were ready to "pull the plug." The original product developed by Ascend had insufficient market, and, in a very high-risk maneuver, Rob convinced his coworkers to change direction and develop an entirely new product. That product proved to be the basis of Ascend's spectacular growth. Rob credits what they learned from this early crisis as instrumental to their later business success. As a result, he has devoted himself to teaching this and the many other business lessons he learned the hard way, at Entrepreneur America and as a frequent lecturer at Cornell University, his alma mater. He now gives a highly regarded class in venture start-ups at Cornell's Johnson Graduate School of Management in which the content of this book is discussed.

Also not widely known, except to those that have been close to him at Ascend and at Entrepreneur America, is the key role of his wife, Terry, in advising him on business matters. Their collaboration is rather exceptional in the world of entrepreneurship, which frequently sees terrible strains on marriages.

The reader will find the style of the book highly personal and intuitive. Each of the many instructive comments is based on the actual experiences of Rob Ryan and others. The overall effect is to give insight and instruction about the creative management skills necessary for a successful highgrowth business. In its creativity, his book can be compared to David Ogilvy's classic *Confessions of an Advertising Man*.

Rob Ryan starts his exposition by analyzing the distinct types of teams that have approached him for training at Entrepreneur America. He puts them in seven "Wanna-Be" categories and insightfully describes the strengths and weaknesses of each, leading to discussion of entrepreneurial core values and what it takes to be ready to talk to possible investors.

Demonstrating an understanding of the market is a prerequisite for establishing the viability of a start-up to potential investors. Rob has a unique checklist for doing this. Furthermore, he provides some sharp advice on getting feedback from potential customers—finding out "whether the dogs actually like the dog food."

Rob then addresses the establishment of the start-up's core competency, the potential for development of a cohesive family of products. He employs a "Sunflower Model" and I, personally, have seen this model stimulate a high level of product creativity in start-up teams.

This, in turn, leads to a discussion of four basic attributes of proposed products: value, differentiation, scaling, and stickiness. The former two are associated with market penetration and the latter two with growth potential. Rob shows how these attributes can be demonstrated.

Next, Rob gives us a very lucid exposition of the art of presentations to investors. Using his own presentations for financing Ascend Communications as an example, he shows, slide by slide, how to build the case for investment. He then discusses the realistic management of the campaign to nail down the investment.

Finally, in the last chapters of this remarkable book, Rob describes the management and organization of a company designed for rapid market penetration, competitive superiority, and high growth. Incidentally, these lessons are applicable not only to start-ups but to new product divisions of larger companies as well.

Throughout the book, Rob includes the exercises he uses at Roaring Lion Ranch to reinforce the important lessons of each chapter.

Rob has tried to spread the word, from his university lectures and his hands-on advice at Entrepreneur America, to as many high-tech entrepreneurs as will listen. They are well advised to listen carefully.

—David J. BenDaniel
Berens Professor of Entrepreneurship
Cornell University, Ithaca, NY

Preface to the Cornell Paperbacks Edition

In Yiddish, the term is *saichel*. On the street, it would be called street smarts. I call it *smartups*—the common sense that produces uncommon profits. The idea of smartups grew out of my career as founder and CEO of Ascend Communications and my experience with Entrepreneur America, the boot camp and mentoring program that I run from my ranch in Montana. Whether I was talking to entrepreneurs from high-tech software, hardware, and dot.com businesses or from low-tech catering and audio businesses, I realized that there's a core set of business laws—common sense—that most startups just don't get. I wrote this book to share those common-sense lessons with all the startups that will never get to boot camp in Montana.

Since the first edition of this book, thousands of Internet "dumb-ups" have been flushed, the Internet captains of industry have been transformed into Lilliputian midgets, and over one million wanna-be entrepreneurs have taken to the street looking for jobs. The sour economy has made it more important for small, medium, and large businesses to become smartups. But the vast majority of companies are dumb-ups. Dumb-ups invent their own rules and ignore the basic laws of business. Smartups practice the laws of business: laws about talking with customers, understanding the customers' applications, creating a strong value proposition, creating differentiation, and many others. Whether your business is big or small, high tech, low tech, or no tech, you can't ignore these basic laws.

They work. I know it from helping technology-based Silicon Valley businesses and low-tech, no-tech small businesses in my own local community of Hamilton, Montana. For example, when we moved here, one of the luxuries we allowed ourselves was a cook (because neither my wife nor I ever liked doing dishes). After a succession of hirings, firings, and quittings, we realized we were incompatible with the genre known as chefs. By then we had met Chris and Terri Daniel, the owners of a catering business, Absolute Vision. Chris and Terri are very good in food preparation and even better with people (customer service). As we became friends over the course of a year, Chris and I began to explore ways to expand Absolute Vision's business. It was doing well, but Chris and Terri wanted to grow faster.

We began what I call a sunflower analysis, one of the smartup strategies discussed in the book. In the Sunflower Model, core competency is everything. Core competency is not a product and it's not a marketing statement. Stripped down, it's tangible, something you do better than anyone else. When we sat down at the kitchen table to reinvent Absolute Vision, the first thing we realized was that its core competencies were food preparation, execution, and service, plus its growing list of happy, well-heeled customers. Given that core, what could the petals of the sunflower (the new product/business/market) be for Absolute Vision? What would best leverage Absolute Vision's core competencies? The first idea was a restaurant. We also looked at a commissary, a place where people could buy well-prepared take-out food. When we analyzed the two petals, it was clear that the commissary leveraged Absolute Vision's strengths and its list of satisfied catering customers. Many of these high-end customers own second homes in Montana and definitely are not interested in cooking but are very interested in good food, wine, and cheeses. Terri and Chris launched their commissary in April 2002 with a grand party. Over a thousand people showed up (who can resist good free food and wine?). The business is off to a great start. The Sunflower Model of leveraging their core strengths and customers served Absolute Vision well.

Here's another example of smartup strategies working for a different kind of business. When I first moved to Montana, I needed a satellite to watch the 49ers football games. Wouldn't a direct TV link be better? My wife and I turned to the yellow pages and hired a company in nearby Missoula to install direct TV. After many glitches and orange-colored wires running to and fro, we fired the outfit. Next, we hired a group out of California not only to install direct TV but also to wire the whole log home for sound. Several months later, my patience expired; the system worked like screeching brakes on a train. Along came Joe McLean of McLean Electronics, recommended by my street-smart Montana ranch manager, Chad Ralls. Within hours, Joe had everything working. We gave Joe more jobs, such as wiring and automating an observatory and installing a media room with large-screen format, video conferencing, Internet. He brought in fifty pairs of twisted phone wires—basically the whole electronics store.

Before he came out to our ranch, Joe's business was installing car stereos. Now he's installed an intergalactic array of sound systems at high-end second homes around the Bitterroot Valley. How did he leverage his core competencies, electronics know-how and customer-service smarts, into adjacent markets to grow a bigger business than a car stereo business?

First we discussed leveraging his work for me into the high-end, second-home market created by CEOs and senior management people from high tech, the financial industry, manufacturing, and retailing. Joe focused on this market and was quite successful. Then we talked about where else we could leverage his core competencies. Joe decided to enter the movie-theater industry; there was competition, but Joe felt his new theater-and-sound system would enjoy an advantage. My sunflower analysis drew a different conclusion: Joe's extensive customer list gave him an opening into the computer hardware and software and network installation business. Joe had an additional core competency apart from his audio talents and that was his base of happy customers. All of those people buying high-end audio for their homes want computer setups, too.

McLean Electronics and Absolute Vision are still works in progress, but both companies are great examples of smartups in the low-tech world. The Sunflower Model is just one of the strategies that gave them the advantage. Leveraging what you do best may seem like common sense, but so often companies ignore the basics. Is there a real business need for your idea, as tested by real customers? Does your business idea involve customers who have money? Does your idea offer value for the customer? How can you grow by competing in other markets? Answering these questions and following the lessons in this book can be the difference between a smartup and a dumbup, no matter what the size. *Smartups* is not just for future $1 billion businesses. It's for any company that wants to reinvent itself and grow to its potential.

Acknowledgments

First and foremost I thank my wife, Terry. She put up with me while I initially wrote this book, reading and editing each revision. She also came up with the new title and cover idea for the paperback edition and helped me to revise its preface.

As with the first edition, it continues to be a great high to work with all of the entrepreneurs. I thank them and now add my graduate students to the list.

Thanks also go:

To Dick Green (Greensy) and his wife, Esme, for putting me up and putting up with me during my visits to our Boston office.

To Sandy Miller, for her good work at our Entrepreneur America office in Boston, including her work on the contract for this book.

To Leah Higgins, my administrative assistant, gatekeeper and right hand; Chad Ralls and Ty Maxey, our street-smart ranch manager and his assistant who keep our Montana boot camp ranch in showcase condition; Dana Nelson, my wife's assistant, who keeps everyone from going crazy by pitching in when things get very hectic; and Diane Holcomb, whose job is to catch and juggle all the balls that come flying in and out whenever we are in Hawaii.

To my Cornell University colleagues, David BenDaniel, who entrusts me with students in the Johnson Graduate School of Management; Janice Conrad, who handles logistics for my frequent trips to Ithaca; Jon Jaquette, who put together EPE, the Entrepreneurship and Personal Enterprise Program that cuts across all of the colleges at Cornell and now includes alumni as well; and Fran Benson of Cornell University Press, who championed this new edition.

To the memory of our moms, who both died last year. My mom, Mildred Ryan, taught me to be street smart, and my mother-in-law, Marjorie Wehe, taught her daughter to give back to the community (and she in turn taught me).

To my dad, John Ryan, who encouraged me to apply to Cornell, although no one in our family had been to college.

And last, to my father-in-law, Bob Wehe, our only living parent, in whose footsteps I follow by teaching at Cornell.

Smartups

Introduction

It was a gray winter day in the Bitterroot Range of western Montana, but high in the Big Sky Gregg Favalora was weaving his way to Entrepreneur America, a "boot camp" for start-ups. Gregg didn't know what to expect. He had heard from other MIT business competition students, Entrepreneur America veterans, that it could be rough on the ego.

Gregg's dream was to build a three-dimensional display for the PC, one that could do modeling, handle design, or just show killer 3-D entertainment. It had been his dream ever since he was in high school, and Gregg just couldn't shake it. While at Yale, he had built a prototype. At Harvard he was supposed to be working on his Ph.D., but he was dreaming in 3-D.

Gregg's flight arrived late at the Missoula, Montana, airport. His three-thousand-mile saga that had begun in Boston ended in the log cabin guesthouse at my Roaring Lion Ranch, home of the Entrepreneur America program.

I got up early the next morning, at six, to work out and get ready for Gregg. In the rustic conference room, Gregg set up to present his dozens of slides. I leaned back in my chair and fired off my first question: "Why would anyone want your product?"

Before Gregg could answer, I hit him with more: "What is the application?" "What is the value proposition to the customer?" "Who is the customer?" "Is anyone else doing this stuff, and are they successful?"

By that point Gregg's scripted business presentation was in shambles, and he was reeling from the onslaught. But then he regrouped, and that's when I saw his passion for his dream, the fire in his eyes. He started arguing back at me, saying that Actuality, his newly minted company, would win because it could build a better mousetrap.

I was impressed by his recovery but still having none of it. " I don't care," I said. "If there are no dogs to eat the dog food, no customers for the product, who cares what the ingredients are?" We spent the rest of the day slicing through his business plan. Several times Gregg looked ready to bolt out the door.

Later that evening, over dinner at my house, he admitted that the session had helped. "I needed this, I don't have the answers," he said. Gregg was not the first entrepreneur, or the last, to meet with me and submit to my free style of no-holds-barred mentoring, a cornerstone of Entrepreneur America.

Who am I to dish out this kind of advice? Well, I've got a few years of experience under my belt. In 1989 I started a company called Ascend, which ended up being the leading manufacturer of boxes that ISPs use for dial-up Internet connections. Ascend's revenue climbed from $16 million to $1.3 billion in five years. We went public in 1994, when the stock soared more than 700 percent. *Business Week* crowned us the top small public stock of the year. We were the first company to make the Fortune 500 list after only six years in business. Then, in 1999, Ascend was acquired by Lucent Technologies for $22 billion.

From Silicon Valley to Montana

By the time of the Lucent deal, I had left Ascend. About a year after the IPO I had serious back surgery and stepped down from running the company. I toyed with the idea of becoming an angel investor but decided enough people were already doing that. I was in a position to give something back to the business world.

I wanted to do something different. I did not want to do another start-up (been there, done that). I did not want to be spreading one hundred thousand seeds, seeding $100,000 to start-ups like Johnny Appleseed, hoping for a few sprouts to grow into trees. I wanted the challenge of sitting down with all kinds of start-ups—dot.com companies, business-to-business companies, software infrastructure companies, hardware companies, even semiconductor companies. I wanted to share with them my know-how and also learn from them. I wanted to systematically codify "how to" and share it with people through my Web site, www.entrepreneur-america.com, and this book.

I had been talking to a wide variety of start-ups, entrepreneurs contacting me to ask for advice and help. I began to think about what makes one entrepreneur and one business model better than another. What clues about the entrepreneur's core values and character portend greatness or bozo-ness? What steps could be taken to guide an entrepreneur? I began to form a mental checklist for reviewing my start-ups, and that became the basis for the Entrepreneur America program.

It might seem that the best place to implement this dream would be Silicon Valley, exactly where I was after Ascend. Wrong. Silicon Valley has tons of entrepreneurs, tons of angels, archangels, devils, incubators, and rip-offs. I wanted to be different. I wanted Entrepreneur America, my mentoring organization, to be different. Silicon Valley is so full of itself. It is not real. My wife, Terry, and I wanted to live in a real community. A community where the people you meet are not on the make to do a company or become a billionaire, nor, for that matter, do they even know or care about the Internet. A community where Ascend could just as well mean "ass end of a moose." So I decided to start by "leaving the valley" to go to Montana.

At the time, I didn't know it was going to be Montana. In fact, I didn't know where we would land. I retained a ranch broker (in case you are wondering, I was born in the Bronx, New York, and raised on Long Island). I gave him a laundry list of things the ranch needed. It needed to have great mountain views, rivers flowing through it, an airport within forty-five minutes, a real town, not a "foo-foo tourist town" within ten minutes, and so on. Oh, and it had to have a guesthouse for my entrepreneurs and an office for teaching and talking.

Nick, our ranch broker, found two places—one in British Columbia, the other in Montana. We arrived to see our ranch on a blustery day (twenty-seven degrees below zero). It was everything I wanted and we bought it. Much to my wife's surprise, we packed up our California house, sent it by truck, and got in the car to find Montana.

Entrepreneurs contact me through word of mouth, referrals, and on-campus lectures at Cornell, MIT, and Stanford. Generally the contact is e-mail (rryan@eamail.com) or registration on my Web site. Either way the entrepreneur ends up sending an executive summary describing the business. If I think it has promise, I set up a phone call or a meeting in Silicon Valley or Boston. After the meeting, the entrepreneur most likely has a big homework assignment. Once the entrepreneur has completed most of the assignment, the start-up team is invited up for a working session at the ranch.

They make their own way up, flying into Missoula, Montana. They rent a car and drive forty to fifty miles to the ranch. When they get there, I get them settled into the guesthouse. Generally the rest of the day (and probably night) is free. The following morning at 9:00 A.M., we dive right into our first session, and we don't quit until the team drives away a few days later.

The Road to Ascend

Compared with a lot of people I've mentored at the ranch, I was a slow starter in the world of entrepreneurship. I wasn't one of those kids who ran a high-margin chain of lemonade stands when barely out of kindergarten. I entered the work world and spent my first ten years there employed by big companies like Burroughs, Digital, and Intel, helping to design new products.

I worked with some great people, but night after night I would come home and complain to my wife, Terry, about how I could do things better on my own. Then, in 1983, Terry came home with one of the first models of a home computer. "What's this for?" I asked. "It's so you can work on your business plan," she said. Later that year I launched a little company called Softcom. It was based on a great idea for making Ethernet cards, which are little devices that let computers "talk" to each other. Softcom ran out of money before we could build the product. That short, disappointing experience taught me two big lessons: 1) Never start a company without a first-class team; and 2) Make sure you've got enough money.

My Softcom experience didn't sour me on being an entrepreneur. Just the opposite; it whetted my appetite. I went back to work for a company called Hayes, but by 1989 I was unhappy again. One day I simply lost it and started packing up my office. Three of my co-workers decided to join me.

In my previous start-up, I had a great idea but no team, no plan, and therefore no chance. This time I was determined to organize my team, write a business plan, convert the business plan to a twenty-minute business presentation, practice the presentation as a team, and aim ourselves at top-tier venture people. In short, I would begin the company the right way by doing all the things I didn't do with Softcom.

Armed with a seventy-page business plan, a thirty-minute presentation, and loads of practice, we contacted our first venture capitalist (referred by a friend), Burr, Egan, Deleage & Co. in San Francisco.

With loads of optimism we arrived fifteen minutes early at the venture capital firm's downtown San Francisco offices. We delivered our presentation, then asked for the investment. A partner named Mr. Deleage leaned forward and said, "We love the team, but could you do something different?"

Stunned, I asked, "You mean you want us to redesign our business model right here, right now?"

His response was quick and angry: "Are you mocking me?" He rose from his chair and left the room with a thud from the door.

"Well," I said, "I guess we don't have the deal." The room erupted into laughter. After a few minutes, in which the associates and others congratulated us on one of the most interesting meetings ever, we departed.

Trudging back to our office, we felt that we'd better have some story for what just happened. A "no" from one firm tends to earn more of the same from others. As we discussed the meeting, we realized that we hadn't done our homework on Burr, Egan. If we had, we would have known that they never made any networking company investments. Why would they begin with us?

Two days later we met with the legendary firm of Kleiner Perkins Caufield & Byers. The meeting was set up through Jim Lally, my old boss who had become a firm partner. Minutes before the meeting, we got a phone call. Jim couldn't meet us, he was at the airport about to catch a flight to Boston. What could we do?

No problem. We raced out to the airport, armed with a flipbook and a fifteen-minute version of our presentation. We made the pitch in a crowded Red Carpet Lounge. I asked for the investment, and to our delight Jim said, "I like it, we are going to do it." He took a copy of our plan, read it on the plane, and called that evening confirming his interest.

Impressive start, but we were not finished. Jim couldn't

guarantee that Kleiner, Perkins would do the deal, only that he would sponsor it. A few weeks later we went to the Kleiner, Perkins offices for the "partners' all-hands meeting." We knew that we'd have to explain the mess at Burr, Egan.

Once again, there we were in the posh offices of a top-tier VC firm. Things were not getting off on the right foot. The partners were drifting in one at a time, some seated, some standing. Questions were being fired at us about the business, about our now famous Burr, Egan meeting. I felt we might self-destruct unless I got control. Loudly I announced that we would answer all questions as soon as we started and that we would start when everyone was seated.

Bold move, but I had done exactly the right thing. Jim, our Kleiner, Perkins sponsor, was grinning from ear to ear in the back of the room.

The room was quiet as we introduced ourselves. The first question came from the audience: "What happened at Burr?"

I stood quietly for a moment, then explained, "Burr and company were brain-dead when it came to networking. They have missed several revolutions in networking, and they are going to miss this one, too."

Silence was followed by laughter, followed by a thumbs-up from Jim, still beaming in the back of the room.

Kleiner, Perkins came on board, and Ascend was off the ground. Greylock joined in to give Ascend $2.5 million. In return, we gave up 50 percent of Ascend.

Under our first business plan, Ascend made network equipment to handle data transmissions over ISDN lines. The product worked great, and we managed to get $2.5 million in venture capital funding. Then we had a small problem. The phone companies decided to pull back their ISDN service, leaving us with a device but no lines to connect it to. That left us with some very puzzled and unhappy investors.

So we regrouped and refocused (or, more candidly, grasped at another idea). We decided to make equipment for telephone videoconferencing, selling our cost-saving product to corporations. It did all right, but the market was limited. Our sales in 1993 were only $16 million. That wasn't enough to satisfy our venture capitalists, who by then had given us $19 million and were expecting a return of seven to ten times their investment. So we had to reinvent ourselves again.

Then, in late 1993, we finally got it right. We went out and talked to dozens of Internet service providers (ISPs) and discovered that they had a serious problem in their back rooms. Because the ISP business was doubling every six months, they couldn't buy and install new modems fast enough. Their back rooms looked like exploded spaghetti factories, with wires and modem boxes tangled all over the place.

We realized that they had a problem we could solve. Ascend could create the plumbing of the Internet, taking our videoconferencing boxes with their high-speed phone lines and using them to funnel lots of Internet dial-up calls into one box. The ISPs would save lots of money because they wouldn't have to buy so many modems. The rest is history—and a ton of work.

The Entrepreneur America Program

I wrote this book to cover the entrepreneurial lessons I've learned, the ones I teach at Entrepreneur America. My approach is built on my years of experience negotiating the peaks and valleys of running (and financing) a successful company. It's a carefully staged process that begins with building the proper team and ends with managing your board of investors.

My first step is to size up the Entrepreneurial Wannabes, which I talk about in chapter 1. I try to learn more about which type of entrepreneur you are and just what state your company is in. Inevitably I find that 95 percent of my entrepreneurs are not ready to talk with a venture capitalist. They have not answered some very fundamental business questions.

Next, in Silicon Valley slang, "Do the dogs like the dog food?" This is the most fundamental question, which asks if customers really, really like and need what you are offering. What kinds of things are important about your product? Some, like Virtual Ink's Mimio, have the dogs ripping open the packaging to get one. Others engender a great big yawn. What distinguishes them? Most of my entrepreneurs have some decent technology, but the product, application, and customers targeted are really boring. Chapter 2 will get you positioned so you are aimed at the right target.

Chapter 3, "The Sunflower Model," outlines one of the most important exercises we do at Entrepreneur America. We draw a sunflower on the whiteboard. The center of the sunflower is the company's core competency, the petals are all the possible ways to leverage that core. One petal is the company's existing product and existing application/market. After getting a handle on the sunflower, next up is to ensure that the products and services are positioned in the best way possible.

So you think you know who the customers are? What the application is? What the product consists of? Chapter 4, "The Keys to the Gold Mine," concentrates on the key questions to answer in building a solid business. The first series of questions includes "How do you make money?" "What is the fundamental business proposition?" "Where are you aiming your product?" "Do the customers have money?" "Do they have a severe need for your product?" and "Do they currently spend lots of money on this need?" The second area to examine is, "How much money do you save or make for the customer?" Number three is, "How differentiated are you from competitors?" Number four, "Does your solution spread with no further salesmanship?" Companies that can answer these questions are ready to start "telling the story" to investors.

All the entrepreneurs I meet are looking for money. Many are eager to set up meetings with venture capitalists. In fact, many already have, and a good proportion of those polluted the wells because they were not ready. Entrepreneurs who have done the homework from Chapters 2, 3, and 4 are ready to put it all together. Chapter 5 gives an annotated example of a winning business presentation and executive summary.

Following chapters 1 through 5 puts you on the way to a pretty decent business model. How do you take that decent model and turn it into a real home run? Chapter 6, "Sucking the Air out of the Room" is all about how you become number one and stay number one. It will tell you how to make selling against your company a nightmare.

Having raised money from VCs or angels, it is time to get on with leading and managing the company's growth. Chapter 7, "So You've Got the Money, Now What?," guides start-ups on how to manage operations, hiring, and the board.

You will read about how my start-ups, each in its own way, were guided by these lessons. You will watch companies like LookSmart move within eight months from impending disaster (unable to meet payroll) to a successful initial public offering (IPO) with Goldman, Sachs. You will see Silicon Spice, a dream company from two young MIT entrepreneurs, raise over $60 million and build a world-class management team. Silicon Spice was bought by Broadcom for over $1 billion. You will read about the inspiring story of RightNow Technologies, which grew without venture capital. RightNow did it the old-fashioned way, by building a product, selling it, and growing on cash flow. That company epitomizes bootstrapping, and they did it in

Montana. As I write this, RightNow is currently going public with Credit Suisse First Boston.

Virtual Ink, conceived by four young MIT engineers, is another amazing story. They took something as mundane as the conference whiteboard and turned themselves into an exciting IPO-destined company. Virtmed looked at the health information field and came up with a brilliant idea of how to help clinics and hospitals with their billing. Virtmed is a classic example of following the lessons. See how they staged their funding and are building a powerhouse company.

These are just some of my Entrepreneur America companies. Since my Ascend days, I have seen over fifty startups at the ranch. I have agreed to go on the board of about a dozen. The lessons I've learned, and applied, can help you, too. When you've finished reading my book, you'll have a solid understanding of how to raise money and build a successful business. Now let's get to it!

1

Which Wanna-be Are You?

ROAD MAP: ▶ Slow down! You're not ready to talk to investors until you've got answers to the probing questions they're going to ask. Take a deep breath and then take stock of where your company stands. You've got a lot of homework to do before shopping for money.

One hot summer day a few years ago, Steve Hau and his start-up team arrived at my room in the Four Seasons hotel in Boston. Steve wore that hungry, desperate look that says "I need money now."

After little introduction, Steve began his laptop presentation as I leaned back in my chair. Steve outlined the vision that he had dropped out of a Harvard Ph.D. program to pursue: "Clinicians don't really have access to information. For example, doctors are still using index cards to capture inpatient billing charges, i.e., the hospital's most precious financial information! A week or two later, those doctors submit the cards to accounting," he explained. "Some 3 to 7 percent get lost, the rest are entered into a mainframe by a pool of error-prone clerks."

Instead Steve wanted the doctors to carry handheld computers (like Palm Pilots) with his team's enterprise-enabled software, enter the charge on the Palm, then place the handheld on an electronic synchronization cradle to automatically send the data to the mainframe.

Steve posed, "Why is it that the Avis guy can zap your rental car with a handheld device and have your life story at the point of sale . . . but your doctor cannot access the most basic information about his patient at the point of care?"

Breathless, Steve finished his presentation. The team's eyes were on me as Steve pushed toward his finale, saying, "We need money. I am three weeks from closing our doors. Can you help my team?"

I sat up in my chair. "Do you have a customer?" I asked him. "Do you have a working prototype?"

Steve mumbled, "Well, no, we don't have the money to do all that."

"You wouldn't get any, either," I said. "Not from me or any top investment firm until you do."

It's the same advice I give to almost every start-up that pitches me. They usually think they're ready for investors. In reality, they generally have months of hard work ahead of them. Over and over I tell them the same thing: "Until you have product and customers, you aren't ready to raise money with top-tier venture capital firms, and those are the only ones I deal with."

Steve looked deflated. His idea had promise, and I offered some advice.

"You need to get some friendly angels, like Mom or Dad or good friends," I said. "Raise two hundred thousand dollars or so to finish your prototype and then get a beta customer, one that will test the product."

Many start-ups just roll their eyes when I shell out this tough assignment. But Steve rose to the challenge. He asked, "If we do this, will you help us?"

"Yep, I'll invite you to Entrepreneur America at my Montana ranch," I said. "We'll work on your business plan and polish your investor presentation."

I wasn't sure what to expect. I give the same advice to a lot of start-ups, and most I never see again. They just give up. But several months later, I was lecturing at MIT. After the talk, as I did the business card shuffle with audience members, I spotted Steve standing at the edge of the crowd.

"Do you remember me and my idea?" he asked. I did and asked him what progress he had made.

"I raised the money, built the prototype, grew the team to five engineers, and we are currently testing our product with a dozen doctors at a large Boston-area hospital," he reported. "So, can my team and I come up to the ranch?"

You bet.

Diagnosing Wanna-be Madness

I've worked with dozens of start-ups at Entrepreneur America. That's the mentoring program I began shortly after leaving Ascend, the Silicon Valley wide-area-networking company I founded in early 1989. In running Entrepreneur America, I've seen all types of company founders. Geniuses, bozos, wonder kids, tricksters, you name it. Lots of people have a germ of an idea kicking around in their heads and are convinced they can turn it into a gazillion-dollar business. I call them "Wanna-bes." I don't use the term to be pejorative. Often Wanna-bes can transform themselves into successful entrepreneurs—but only if they're willing to work hard.

The classic error that a lot of my Wanna-bes make is mistaking the *idea* of a business for the actual *building* of a business. By coming up with a good idea, they feel they have already done the hard part of building a company. In fact, what they have done is equivalent to finding their sneakers before running a marathon—they're still not even at the starting line.

Most haven't done their homework on the business model. They haven't built a prototype or gotten feedback from potential customers. Frankly, very few are ready to raise money. My main mission with these entrepreneurs is to slow them down and get them to start asking (and answering) fundamental questions about their business.

Not only do I teach the entrepreneurs, but I learn from

them. One thing I've noticed after four years of working with start-ups is that there are a few distinct types of teams. The philosophies and approaches the founders take toward building companies tend to place their companies in one of seven categories.

- ▶ Quickie
- ▶ Wonderful Wacky MBA
- ▶ Send Money
- ▶ Dreamers
- ▶ One-Stripe Zebra
- ▶ Technoid
- ▶ Guts and Brains (the Dream Team)

All these categories, except Guts and Brains, have one thing in common. They have not done their homework on their business model. They are not ready to raise money. My main mission with these entrepreneurs is to slow them down, strengthen the team, and get them to start asking and answering fundamental questions about their business.

THE QUICKIE WANNA-BE

In the Internet economy of the late 1990s, Quickies have been popping up like jackrabbits. Quickies are identified by a get-rich-quick business model that has no clear-cut application and no value proposition. These money-losing business models just don't make sense for building stability, for growing a sustainable company that will last for ten or twenty years.

One classic profile I've seen over a dozen times (but don't mentor) is the "Eyeballs.com Ponzi scheme." The founders either talk about amassing consumer "eyeballs" to look at a set of content or use a marginally helpful product. Making money is not a concern. In fact, it usually costs more to attract the eyeballs than the company earns from getting it. Or, in many cases, the company just gives away content or product. The idea is to amass the eyeballs, go public at the earliest opportunity, manage the stock for maximum valuation, then sell the company before the six-month holding period for stocks of insiders has expired. Voilà! Magic wealth for the founders, VCs, and insiders.

Whether the market condones this or not, it's not okay in my book. This approach doesn't build anything of value and benefits only a small number of people. Quickies who actually want to build a company need to focus more on growing the business and less on managing a stock. Real companies build predictable revenues and large, profitable earnings. Wealth creation follows real companies. I can't tell you how many people have told me that profits from Ascend stock sent their kid through college or paid for a new addition on their house. That feels good because not only am I benefiting, but they are, too. Who benefits from a Ponzi scheme except the guy who starts it?

THE WONDERFUL WACKY MBA WANNA-BE

I know I have budding MBA Wanna-bes when the "wacky attack" starts. They dance around the conference room, whipping out tons of charts and quotes to prove that the market is humongous. I call it proof of the "zero-billion-dollar market."

Most MBA Wanna-bes are like the team I met a few years ago back east. They were led by a very enthusiastic entrepreneur whose dream was to create a Web site for alumni of major universities. The idea was to create a "place for people who shared something in common" that would offer news, health, and shopping.

The team came to see me and went into a complete wacky attack. They started off with marketing charts that demonstrated things like how many alumni were floating around (millions!) and how much money they spend (zillions!). Like most wacky attacks, this one never got around to important data like exactly who the customer was, what he was buying from them, and what the value proposition was.

Here's what's wrong with marketing reports: Either the industry is immature, in which case nobody knows what the hell they're talking about, or the industry is mature, which means that there are entrenched leaders.

THE SEND MONEY WANNA-BE

Frankly, lots of entrepreneurs start off in this category. You'd be amazed at how many people ask me—a complete stranger!—to give them a few million dollars. They think: Mentor equals money.

These Wanna-bes think that once they get a big VC check in the bank, everything else will fall into place. In fact, it's the other way around. You need to get everything in place, or at least a lot of things in place, before you can start asking for money. Money follows those who do the right things.

For me, the Send Money alarm goes off when I ask, "What's your financial status?" and the answer is, "We're broke, but we just need $250,000 to get over the next hump." Hah. When I ask a few more questions I usually find out they shot their business plan at every VC guy within range, but because they hadn't done their home-work, everybody dodged their presentation and turned them down.

Just last week, for example, I got a classic Send Money letter from someone I didn't know. The letter wasn't even addressed to me, it was "To Whom It May Concern." Here's the first paragraph:

> I have a wonderful project with plenty of long-term earn-
> ing potential. My dilemma is that I lack the initial funding
> to really make this project go. I am asking you to guide me

in the proper direction to fulfill my dream. I am providing some baseline information regarding the business for which I am interested in acquiring financial assistance and guidance.

Classic. After a few more paragraphs outlining a vague business plan, a shaky management team, and a few potential distribution relationships, the letter wraps up like this:

What we are seeking is seed money to create a comfort level so that all the critical elements needed for the success of this venture are able to run smoothly without having to function at so lean a position during start-up of the business in 2000.

I guess it never occurred to this company to bootstrap during the start-up phase. When I get a letter like this, I don't even bother reading the whole thing. This team isn't ready until they demonstrate that their focus is on running around talking to customers instead of investors.

THE DREAMER WANNA-BE

These Wanna-bes are "Visionaries" with a capital V, rarely blessed with detailed information or management know-how. They're grand thinkers and schemers who tend to slip easily into Send Money mode. If you're in this category, you could admittedly be the next Michael Dell or Bill Gates. But it's more likely you'll end up tinkering in your garage forever.

The problem with visionaries is that most of the time they aren't doers. They love to sit in a room and think about their great idea or spend hours telling their bored friends all about it. But they don't know how to snap out of Dreamer mode and turn their glorious idea into something real. Even if they do, big dreams usually come with big problems, and these Wanna-bes aren't always the type to figure out how to work around them.

Let me give you an example. I met a group that had origins at a major West Coast university, a super hotbed of entrepreneurship and excellence. The people on the team ran the gamut, from a CEO who was also involved in fourteen or fifteen other companies (he really wanted to be a venture capitalist but didn't have the experience), to a scientist who wanted to change the air we breathe, to a decent inventor who was relatively time-tested and savvy.

The business idea was to create a new type of cleaner that removed carbon-based substances from the air. It was a fairly clever design for intensifying and burning off the bad stuff. But remember, ideas are like belly buttons: everyone has one.

The real problem with these Dreamers was the team. There was no core, no center of reality. The CEO was superficially involved in tons of other deals, the scientist's vision was clouded by the need to save the world, and the inventor just wanted to make money, preferably in a way that didn't interrupt his work at the university.

The group fought, literally, which eventually disintegrated into a lawsuit. They did actually build a prototype and then tested it on wood-burning stoves. It worked really well, but my reaction wasn't positive. I had discovered that the market was small and the device cost more than the stove.

The team started exploring the muffler business. At first that sounded better, especially after the company received an offer from a knight in shining armor—one of the trucking companies.

But the offer locked the Wanna-bes into that trucking company, and not even for real money, only the *promise* of money. Here's where the pugilistic team dynamic kicked in. The CEO didn't like it, the scientist saw it as their salvation, and the inventor . . . well, who knows? My last glance, as I walked away from the whole thing, was of the CEO getting pummeled by the inventor. Later I heard that the team agreed to the vague truck company terms.

THE ONE-STRIPE ZEBRA WANNA-BE

This is another one I see a lot in Internet start-ups. The One-Stripe Zebra is a company that's built around a single function with a very narrow market. Perhaps it's even an interesting and viable product, but it's just not a wide enough stripe to support a sustainable company.

Instead the product is really part of a bigger story. Usually that story belongs to a larger company, a competitor. The big company likely has a whole product line with which the Zebra's dovetails. Either that big company is already working on developing the Zebra product, or they'll buy it from someone else. Even if the Zebra gets the product to market, it's tough to compete against big guys offering a whole sweep of products. So the Zebra faces competition from the top (big players) and also from the bottom (other start-ups).

Let me give you an example from low-tech world. An inventor came to me with a mechanical device to perform CPR. CPR is very effective in restarting a heart if done within a few minutes of a heart attack. But hospitals (with electrical CPR systems) are usually more than a few minutes away. This mechanical device, which people could keep in their homes, worked by creating leverage to restore the proper pressure and heartbeat for effective CPR. I thought it was a great product; every home should have at least one. Sort of like a fire extinguisher.

It was a very good idea, but still a One-Striped Zebra. The inventor had gotten very focused on the product, which really was terrific. But he hadn't stopped to think beyond that, like about whether or not he could build a company around it. A single product with a limited market might support a small office/home office business, but it probably wouldn't generate enough sales to grow the infrastructure of a major corporation.

I recommended that he talk with companies that had a natural synergy with their other products and could afford to develop the product through clinical trials. Why invest all that time and money into building a company to do it himself? Let the other company's economies of scale propel the product through the trials and distribution. He could just sit back and collect a royalty check every month. The last time I heard from the founder, that's what he was doing.

The Internet company Hotmail was also a One-Stripe Zebra. Hotmail built a product that allowed subscribers to check their e-mail by logging on to a Web site. This was an interesting and useful function, but how the company made money was not clear. Obviously a product like that is more useful as part of a larger operation. In the case of Hotmail, that larger operation became Microsoft, which bought Hotmail a few years ago.

THE TECHNOID WANNA-BE

At Entrepreneur America, the Technoids represent the dominant species. I must have helped at least five MIT start-ups, including Silicon Spice, Virtual Ink, Iridigm, Virtmed, and Actuality. I'll admit I've got a bias toward engineers. Because I was trained as a mathematical logician and spent the first ten years of my career as an engineer at companies like Intel and DEC, I like really smart technical types. The Technoid Wanna-be has a good beginning with a beefy engineering team that can usually build a decent product.

But first they have to overcome a few problems. Technoids are smart about their technology but not always clued in about how to run a business. Because of this, Technoids tend to be vulnerable to marketing consultant scams or "I will raise money for you" scams.

One extreme example of a Technoid team I recently saw comprised over twenty "rocket scientists." They had created incredible Web sites to build community and encourage collaboration—the users were interacting with the sites so much that they were basically generating the great content.

They were a freeware company, which meant that the software was given away. Revenue came from charging clients who wanted the company's expertise on using the software. For that, clients paid $1 million to $2 million a year. Obviously the company didn't need many customers to be a successful revenue-generating company. So far so good on the business model.

The problem was that the company had, at that stage, no administrative team, no sales or marketing people. It had engineers, and lots of them. In fact, when I met them they were hiring an accountant, but the accountant had to be able to code. Not all Technoids are this hard-core about the engineer thing. But given the strong bias of this particular founder, I was sure that even the receptionist answering the phones was probably going to be an engineer.

The company was a loose confederation guided by the founder and his co-founder. When I looked at their business model I realized three things. It was good, but they didn't have the business people to handle the management issues.

He had promised the rest of the engineers that they would be able to cash in their stock at some time in the future. But he didn't know how to do this while also keeping tight management control over his company.

After I met with him briefly, I pointed out a few key things. There were several potential companies that could spin out of his technology and, with some hard work and a little luck, eventually go public. But it would take time and several milestones before that blessed event. First, we needed a predictable revenue model. Second, we needed a management team with a track record of working together at that company.

I ended up drifting away from this team. I believe it

could have been fun, but only if the founder were a little less Technoid.

THE GUTS AND BRAINS WANNA-BE

These are the guys and gals I like. These are the ones who make it, and the main reason they do is because they're smart, but not just book smart. They've got the guts to plunge into the real world, even though there is a lot of scary stuff out there. They dig in and do their homework. Sure, they'll get rejected and ignored at first, just like all the other Wanna-bes. But this team keeps going. They've got faith in themselves and their business.

An example is one of the start-ups I got closely involved in at the ranch, a company called Actuality. It was founded by Gregg Favalora, a young Yale- and Harvard-trained engineer. His passion is to build three-dimensional volume-filling devices so that you and I can see things on his hardware device in 3-D—*real* 3-D, not 3-D on a two-dimensional screen. It's sort of like the Princess Leia animated hologram in the first *Star Wars* movie.

This device caught my imagination immediately because it had several promising applications. It could be used for modeling and design for any kind of product development. It has clear possibilities in the entertainment industry—3-D e-mails or Web video sprang to mind. There is potential in Web commerce and on-line shopping.

I first met Gregg at MIT, where he and a small group were the business corporation winners at one of the Sloan School of Management's start-up pitch contests. My advice to Gregg was to find some excellent software people to work with—developers to build a prototype. Then he would have to beta-test it in a few different markets, to learn which applications had the strongest potential for development and growth. I wanted him to patent the technology.

Because the company was broke, Gregg could have easily slipped into Send Money mode, but he didn't. Partly it was because I helped him stay focused, but mostly it was because he has the right stuff, and he trusted his inner voice.

He used his savings, money from his parents and friends, and moved into a cheap basement apartment. He found like-minded entrepreneurs, and they worked on the prototype. They "bootstrapped" by scrounging bits and pieces—for example, the light sources for painting the image were ordinary laser pens from an office supply shop.

Gregg handled the marketing himself by getting on the phone and being persistent but friendly, inviting people at all sorts of companies to either visit his apartment or let him take the demo to them. Most came to him, sending three or four big company hotshots to Gregg's student-looking basement rental in Cambridge, where a pot of coffee and box of doughnuts sat on the makeshift table, next to Gregg's little flipbook presentation.

What Gregg did right in his presentation was to rehearse it relentlessly. Because of that practice, when it came time for the real thing he was confident, but not cocky. He used a simple flipbook presentation, then showed his visitors the demos of a flying helicopter and beating heart.

Well, who wouldn't be interested? The visitors liked Gregg's team and preparation but had several concerns about investing in the 3-D market. Could Actuality build the product at a reasonable cost? Would the team stay together and attract top management? Was there a driving application market? I believed the answer to all of these was "yes," but for over a year Actuality's team ate rice and beans while trouping around to VCs and potential investors. No dice.

Then, for a minute, it looked as though Actuality had hit the big time. A former VP from Silicon Graphics agreed to be the CEO. But after three months of beating his head against a wall looking for funding, the CEO quit for an easier job. After this disappointment, who could blame a Wanna-be for giving up? But Gregg didn't. Still eating rice and beans, he started targeting angels.

Finally, after two and a half years, Actuality had a real breakthrough. Several VC firms and angels ponied up $600,000, which quickly swelled to over $1 million, enough to build a working model of the product. Gregg embodies Guts and Brains; he'll see his vision through, and I'm proud to be a part of it.

Core Values of an Entrepreneur

In 1997 a start-up team visited me in my home in the San Francisco Bay Area. As we sat around the pool, the two co-founders described their idea for using the Web to let people sell collectibles like Beanie Babies through on-line auctions.

My first thought was, What the hell is a Beanie Baby? My second thought was that although the idea sounded intriguing, I didn't see enough barriers to entry. How could the company eke out a profit on such small transactions? The start-up was looking for money, and the MBA and his engineer partner offered me 5-10 percent of the company in return. I wasn't thinking of myself as an angel investor, and I turned them down.

Well, let me just say that it was one of my worst bonehead moves. The company is called eBay, and if I had taken that stake from Pierre Omidyar, it would be worth buckets of money today. Oh, well. That was my first investment opportunity since leaving Ascend, and I still had a lot to learn about judging start-ups.

The eBay experience made me start to study what makes one entrepreneur and one business model better than another. What clues portend greatness or bozo-ness? I began to form a mental checklist for reviewing my start-ups. First on that list is the character of the founder and whether or not he or she has the core values necessary to succeed.

The born entrepreneur comes from a rare breed of people with intelligence, great heart, and creative skills. They're visionary and self-confident; good communicators with unlimited energy and a passion for what they do. They are prepared to endure lean times and make sacrifices, relying on strong willpower and competitiveness. A true entrepreneur doesn't see change as something to fear or even merely have tolerance for change. Instead he or she has an appetite for it, recognizing that change is what brings opportunity.

An entrepreneur knows how to build a strong team, balancing his or her own weaknesses with others' strengths. The team is the first company element that investors assess. They know that a hardy team can overcome staggering obstacles, quickly whipping up new products, outsmarting competition, recovering from problems, and making impressive investor presentations. The team, not the product, is the root from which the entire company grows.

Some people are born with the right entrepreneurial instincts for assembling a great team, some aren't. I'll tell you flat out if you just don't have the skills or even the potential to learn them. There's no sense in wasting anyone's time. Fortunately, for those who weren't born blessed, I've discovered that most solid entrepreneurial traits can be learned once you know where to focus.

An entrepreneur must be **intelligent**—bright and quick enough to grasp concepts, implement visions, and cut through distractions to focus on important issues. As one of my Ascend board members, a prominent venture capitalist and entrepreneur, said, "Rob, if the people don't have the IQ points, I don't think they are going to grow any during the start-up." Admittedly this one is hard to nurture. You're either smart or you're not.

Initiative is key, and is something that can be learned.

Some entrepreneurs have to be coaxed out of the back room. They are more comfortable sitting with their colleagues and discussing abstract business models than going out to customers and doing the work with shoe leather and sweat to see whether the theories hold any water. That's a lack of initiative, and those people won't make it.

Integrity is non-negotiable, because business *is* relationships. If people can't trust a company founder because he or she changes his mind, twists things around, or is always looking out for number one, then it will be impossible to grow a great company.

It's an intangible item, but **heart** is indispensable. Heart bonds people together. It is what makes me realize that a start-up founder is driven to get things right. He or she is willing to show weaknesses, trusting and believing in people. Sure, sometimes an entrepreneur with heart gets burned, but most of the time that trust is returned in full measure.

Successful entrepreneurs have **willpower**, enough to burst through barriers. Instead of being overwhelmed by problems, successful entrepreneurs tend to take the perspective of, "I can overcome this, there's no way we are going to fail." If that attitude is strong enough, it can carry a whole company and sweep up investors as well.

From Wanna-be to Working Company

One of the first entrepreneur teams I invited out to Entrepreneur America was a start-up called Creditland.com. I had met the founders at my house outside San Francisco. They wrangled an introduction from a friend of mine and came out to the house to show me their business plan.

As I sat through the first few slides of their presentation I realized that they had great potential. The idea was to create a Web site where consumers could apply for a credit card, auto loan, or even a home loan, choosing from dozens of financial institutions like MBNA and Bank of America.

Five minutes into the presentation I invited them to Entrepreneur America. I could see that they had the kernel of an idea but needed help with the execution. I knew I could point them in the right direction.

Ten days later, after some schedule juggling, the team showed up at Entrepreneur America. I call it my "boot camp" for entrepreneurs because of all the hard work we do. The Creditland team liked the metaphor—all four of them, three men and one woman, were dressed from head to toe in military fatigues.

It's a good thing they were so enthusiastic, because we had plenty of work to do for the team to move from Dreamer Wanna-be status to the Guts and Brains level. Over the next two days (still wearing the fatigues) we polished the business presentation and focused on setting up solid team operations and building competitive barriers. I worked with them to identify the missing people for the organization. We also created a project tracking system with well-defined goals.

They listened to me and did their homework, and it paid off. Today Creditland.com has signed up several banks as customers and has completed a $15 million round of investing.

It wasn't as easy as it sounds. I worked Creditland hard at the ranch, then sent them home with a mountain of homework assignments. Many lesser teams simply quit, overwhelmed by the amount of work. They were just too lazy to be entrepreneurs. They'd rather dream about checks falling into their laps.

As the chart illustrates, most companies just aren't fundable right away. Even if you're in the Guts and Brains category and already lining up meetings with VCs, it will take one to three months before you're depositing checks.

You Ain't Ready Unless:

1. You have built your product or a prototype or a demo.

2. You have talked to customers and they like it, and they have provided you with references.

3. You have talked to manufacturers and it can be manufactured. (Skip this step if your product is software, or ASP, and so forth).

4. You know how you are going to make $ (buckets of money).

If 1 to 4 are done, you "is ready" or you ain't.

Why do my start-ups and I go to all this trouble? The fast growth of investment money, the velocity of the Internet, it all points to speed. Why not just take the idea and hit the VC trail? There are several hundred in the Sand Hill Road area of Silicon Valley alone. Surely (after forty or so meetings) one investor will bite, plus they'll straighten out your business model once they invest.

Well, I'm not going to say that the above scenario wouldn't work. It *might*, but not with the top-tier VCs I work with. Even landing an investor is a hit-or-miss proposition, and I'd rather gamble with better odds. It's also not the style of someone intent on building a great sustainable company. For that you have to do your homework. The next assignment, after building the team, is to focus on the product development process covered in chapter 2.

The start-ups that follow the approach I teach at Entrepreneur America get funded. Period. That's all the more remarkable because I'm not seeing the cream of the crop out here in Montana. Jim Clark and Steve Jobs don't need me because they've already got investors camped out on their sidewalks. I'm seeing the guys who've been getting only a stale bagel at the VC breakfasts. It doesn't mean they are losers, just that they have a lot more work to do before they're ready to go out and raise money.

Most Wanna-bes are just chasing their tails. This isn't the time for VC trolling, it's the time to chase customers instead and test out your ideas on them. You can't raise money until you understand your business model and do your homework.

EXERCISE:

Symptoms of Wanna-Be Madness

---

1. Did you form a hastily assembled, lopsided team, with lots of "B" players? Was your attitude that it's more important to get bodies on the team than to make sure they are the right ones for the job?

2. Did you pick a really hot industry or niche, based on television and analyst buzz?

3. Are you drawn to the business model of the week (for example, consumer.com)?

4. Do you have a team with three or four MBAs, college buddies, and maybe one junior engineer?

5. Does everyone have significant titles, like CEO, COO, and president?

6. Do you find yourself reading analyst reports and getting excited about major market opportunities and trends?

7. Did you destroy a forest to create your business plan?

8. Does your business slide presentation have lots of market graphs and analyst quotes?

9. Do you have the urge to talk to everyone about your idea? Do you do all the talking?

10. Are you walking around talking to every venture capitalist who will listen to your idea?

11. Do you find yourself asking strangers for money?

12. Are you avoiding bootstrapping?

13. Are you avoiding customers?

14. Are you irritated with the VCs you're talking to?

15. Is your product more like a single function?

16. Does your function really fit better as part of a competitor's larger group of products or systems?

17. Is your company "vision" something intergalactic in scope—vague and unable to be patented?

18. Are you afraid to talk to customers, fearing they might steal your idea?

19. When you explain your business idea, do you have to go into excruciating detail so that they understand?

20. Is your management team top-heavy with engineers?

21. Have you fallen in love with your company's technology?

22. Do you have only a vague idea of where and how to apply your company's technology in the marketplace?

23. Does your management team have a couple of strong engineers and one experienced marketing person?

24. Does the team have years of experience in the chosen field?

25. Are you developing a business hypothesis, then testing it on potential customers?

26. Has the team built a prototype or demo to show to customers?

Answering "yes" puts you in one of the following Wanna-be categories:

1-3: Quickie
4-9: Wonderful Wacky MBA
10-14: Send Money
15-16: One-Stripe Zebra
17-19: Dreamer
20-22: Technoid
23-26: Guts and Brains

▶ You are welcome to refer to the Entrepreneur America Web site for additional information (www.entrepreneur-america.com).

Do the Dogs Like the Dog Food?

ROAD MAP: ▶ You realize that your company isn't quite ready to talk to investors, but what should you be doing? Focus on product development. Make sure it's something that doesn't already exist, that customers are dying to have, and that will make money for your company.

About two years ago I sat in my Boston hotel room, watching a laptop demo from yet another start-up team. I hear over a hundred pitches a year; I can tell pretty quickly whether someone's driving a hot rod or a lemon.

These guys, Netcracker, had a potential hot rod, but their marketing plan was a lemon. What they had was a network simulation package that ran on a PC. The database stored information on types of network equipment from various vendors. In the simulation, a user could select sample equipment from the database to build a model, then see if the parts were compatible and whether or not the whole system would work when actually hooked up.

These guys were Technoids with a capital "T." They had a good database, but no understanding of the business model, product, or customer. The team droned on and on about licensing the software and database to corporations for simulating their in-house networks. I quickly interrupted.

"The database is interesting," I said. "But what you guys are planning to do with it is really boring."

They stared at me as I continued.

"You need to focus on making money from the database right away, not the simulation," I said. "Use the Web to do it." Like telephone companies using it for network equipment inventory. Or a tool for network resellers to use in ordering equipment. Or maybe corporate clients could use it for purchasing and configuring equipment.

But the founders disagreed. They left my hotel room confused and didn't do anything I suggested. Nine months later I got a phone call from the son of one of the founders—he had replaced his father as CEO (Dad became head of research and development). He wanted to go in exactly the direction I had recommended earlier. I invited him out to Entrepreneur America, and we started working together.

Now they've got four potential markets for this product, and they're about to enter into trials with one of the largest network systems distributors and a large telephone company. Netcracker has done its customer homework. By landing $2 million contracts, it has created choices for itself. The company can raise venture capital money or build the company off cash flow.

The story illustrates a common mistake made early in the start-up process—namely, the company may have an interesting core ability or technology but frequently develops it into the wrong product and aims it at the wrong market.

A common question tossed around Silicon Valley venture capitalists is, "Do the dogs like the dog food?" It means, Do customers like the product? Are they drooling to tear open the package? If the answer is "yes" (and I'm going to tell you how to make sure that it is), then you've got investors halfway there.

Rob's Quality Dog Food Test

Rate each of these categories with a number from 1 (worst) to 10 (best):

1. Idea—Is it sufficiently developed to have customers? That's a 10. If it's still in the dream stage, it's a 1.

2. Customers—Is the idea aimed at *paying* customers? Rate their potential from 1 to 10.

3. Money—Is it clear how the business makes money now (not ten years from now)? If it's a very clear, easy-to-explain model, give it a 10.

4. Application—Is there a clear business problem that the product solves? If so, it's a 10. If it's just kind of fun to have, it's a 1.

5 Uniqueness—It ranks a 10 if the solution is unique Give it a 1 if it's merely a variation on some existing theme.

6. Value—Is the value proposition extremely clear? Does it save a measurable amount of money? Then it's a 10.

7. Barriers—Does the product, once built, throw up sufficient barriers to entry? If the barriers buy the company in about a year, it's a 10. Lower the rating for less time.

8. Strategy—Does the product or service directly impact the customers' bottom line? Or is it just nice to have? (For example, customer relations management is strategic—a 10. Training internal people on Office 2000 is important, but not strategic—a 5 or 6.)

9. Competition—Are lots of others doing the same thing? The fewer the competitors, the higher the score.

10. Scalability—Once your customer adopts the product or service, will it spread rapidly to other customers (or other parts of the company, if your client is a corporation)? The faster and more places it spreads, the higher the rating.

Will It Sell?

Most Wanna-bes tend to dream up their business ideas in vacuums, doing very little real-world work. Many rely too heavily on their imagination and abstract studies, interacting too little with potential customers. What they should be doing is checking to see if the product or service idea already exists, if it's radically different from what's already out there, and whether or not customers even care.

The Guts and Brains Wanna-bes figure out how to dish out product that's irresistible. At Entrepreneur America we run all of the ideas through the wringer to make sure the company's product is something customers will buy. We're asking ourselves the same thing the venture capitalists want to know: "Will the dogs like the dog food?" This is my own set of criteria that I use to review every idea that comes my way:

Let's apply Rob's dog food test to a couple of company examples and figure out who's on the right track. The first one is a Wanna-be team I met out in California when I was making a speech, then asked to come out to the ranch. There, we studied their product, a Web-based tool to build three-dimensional worlds that use avatars, or little representative animated people. Basically, it's a chat room on steroids. The company planned to sell yearly subscriptions over the Web to consumers who wanted to play in a really cool fantasy chat room. I labeled this Wanna-be a "Dreaming Zebra"—they had a narrow idea for a business, but it wasn't much past the dream stage.

1. **Idea:** Here was a big problem. Aiming a $20/month product at consumers is a bozo business model. The way I look at it, if you want to steal some money, you rob banks because that's where the money is. You don't rob delicatessens. But these guys, with this business model, were aiming at the delicatessens. Score: 2.

2. Customers: Consumers might pay for it, but they aren't going to pay much. Score: 2.

3. Money: This team wasn't even set up to make money today, much less ten years from today. This plan had no longevity at all. Score: 2.

4. Application: Playing in chat rooms doesn't really solve a core business problem, does it? Score: 0.

5. Uniqueness: Okay, they got me here. I had to admit that the product was pretty unique and well designed. The only thing that kept it from being a 7 or 8 was that the founders didn't actually own the source code. Score: 5.

6. Value: There's totally no value proposition at all. This product just offers fun, because playing around in chat rooms isn't going to save anybody money. Score: 0.

7. Barriers: Even though they didn't own the code, it was fairly safely licensed. They had a fair amount of so phisticated software there. Score: 6.

8. Strategy: From the customer's perspective this product is nice to have, but there's no bottom-line boost. Maybe you can make money on games, but from my perspective, people who just want to play around aren't going to fork over very much money. I prefer business-to-business models. Score: 0.

9. Competition: We didn't have any data on this, since we didn't do the research. If we had gotten a chance to start, we would have talked about who their competitors were. I would have done some Internet research on them. We would have put a matrix together rating their strengths and weaknesses with a financial analysis. Score: N/A.

10. Scalability: This might scale to zillions of other chat room fanatics. Score: 5.

What's wrong here? The people who want to play (consumers) aren't willing to pay. The idea just isn't that compelling, who really cares? The main problem here is the application; there's no real problem this chat room solves.

Because of that, there's no value proposition, no compelling reason for customers to buy the product.

But to me, even with these low initial ratings, this team was worth a second look. The big plus was the appealing 3-D software, which had real possibilities for strong market applications. It looked as though the barriers to entry were reasonably high. So I invited the team out to the ranch.

In Montana I asked them to do a "customer inventory" of who was using the product. We discovered that they actually had a corporate account—Boeing. I asked them what Boeing was doing with the chat rooms, but the founders didn't know. So I put them on a plane out to Boeing headquarters in Seattle to find out. It turned out Boeing was using the product to run virtual corporate training for its airplane mechanics. Aha! Suddenly we had a clue about how to reposition this product and aim it at a rich niche—"3-D e-training."

Here's how the new, improved plan ranked:

1. **Idea:** This was much more promising once we repositioned the 3-D "chat" room as a serious corporate e-training product. Possibly we would have discovered this was a 9 once we tested it. Score: 7–8.

2. **Customers:** Suddenly, with this new corporate training angle, we moved from targeting consumers just playing games to deep-pocket clients like Boeing. Score: 7–8.

3. **Money:** The repositioning aimed the company squarely at the lucrative corporate training market. Zillions of dollars are spent on that. And anyone training people on three-dimensional objects is going to think it's pretty cool. Score: 7–8.

4. **Application:** Now we're talking. This training tool would solve a common and expensive business problem, which is to bring employees physically together to learn new skills. Score: 9.

5. **Uniqueness:** There might be people out there doing

3-D Web design, but training companies still rely mostly on the traditional model of gathering people together physically and in real time. That's the competition, and they just aren't using that technology. My early instincts were that this score was high. We planned to research it further. Score: 9.

6. **Value:** Companies can be more effective if they train on-line instead of shipping people somewhere physically. Suddenly students and instructors from all over the world can meet up on the virtual 3-D shop floor, interacting by using the chat feature. Score: 8.

7. **Barriers:** The barriers get even higher with this new market. In the game market, other developers are more likely to have comparable technology. But low-tech training companies aren't taking advantage of it. What we were doing was moving our technology from the high-tech gaming industry to the low-tech training industry. Score: 8–9.

8. **Strategy:** Training is very strategic for businesses. Labor costs eat companies alive, so anything they can save in that area is a bottom-line treat. Score—9.

9. **Competition:** This would come from traditional training companies upgrading to Web technology, and from newbie high-tech companies using the Web but without much experience in training. The competition would be there, but we hadn't done enough research yet to know how severe it would be. Score: N/A.

10. **Scalability:** As big companies trained their employees, this had potential to spread from one internal application to another and out into the supply chain. Score: 7.

Even though we were heading in the right direction, on track to move from Dreamer into the Technoid Wanna-be category, this story doesn't have a happy ending. The developer (who owned the code) ended up suing and settling with the founders. They were supposed to pay him off, but when money got tight, they missed a payment or two. To earn the money to make it up, they agreed to do a semi-sleazy shell IPO with some investors I didn't like. That's when I bowed out. It's a shame, because I thought the idea had some potential.

I don't have a clue what they're doing now. When these things drop from my radar screen, they really drop. I may get e-mail saying they're the biggest company on Wall Street or that they've broken up the company and gone on to other things, but when I stop working with a company I forget about it.

Here's another test example, one that earned top marks from the very beginning. A start-up I mentioned in the last chapter, called Virtmed, developed an application for doctors and their hospital and clinic billing processes. It's PalmPilot software that uses a few clicks to enter the patent's procedure code automatically, then quickly synchronizes with the hospital or clinic mainframe accounting software. Will the dogs like the dog food? I definitely think so. Will the dogs *buy* the dog food? You bet. Let's rate this one using Rob's dog food test:

1. **Idea:** Virtmed had done some chitchatting about the idea with various people and hospitals. They had a clear idea, but it was still pretty much in the dream stage. Score: 5.

2. **Customers:** The idea was aimed squarely at customers, but Virtmed didn't yet know how "paying" they were. After a little research we would discover that some large hospital accounts are really hard to get money out of (we ended up focusing more on clinics, rehab centers, and so forth). Score: 6–7.

3. **Money:** The future opportunities here were huge. Doctors have a lot to do besides billing, and once we got this tool in their hands it had the potential to handle all of

that. Pharmacy information, patient records, billing, everything that a doctor might want to automate. Score: 9.

4. Application: Doctors want to fix patients, not process bills. Score: 10.

5. Uniqueness: All clear on the immediate horizon. We knew there was one other company trying to do the same thing from the pharmaceutical angle. Eventually one would push into the other's sandbox. But that wouldn't happen for a while. Score: 8.

6. Value: This is high because the application saves money. Right now doctors write up billing information on index cards, and someone has to key them into a computer. Months later the patient gets a bill, so there's a long delay cycle. Also, doctors lose 3 to 6 percent of the cards, therefore losing 3 to 6 percent of revenue. Often they miswrite things on the cards, and when they do that they get sued. Score: 9.

7. Barriers: Once they landed an account or two, these barriers would be huge. That's because it required building an interface to the hospital's existing accounting system, which is a lot of work. There are only about four main accounting systems that hospitals use, and once Virtmed knocked a few of those off, they'd be way ahead of any competitors. Score: 8–9.

8. Strategy: This product is a direct bottom-line booster. It reclaims lost revenues for the hospital, updates cash flow, and saves labor time. Score: 10.

9. Competition: There's no direct competition yet. But obviously the pharmaceutical-angle company loomed on the horizon. Score: 7.

10. Scalability: If the test works with small clinics, they'll want to spread this efficient little operation around. Score: 8.

When I first met Virtmed, I scored them high on their *potential*. We didn't know for sure that the actual value proposition would be that high, but my instincts told me it would pan out. Basically I was scoring the hypothesis, then we had to go out and test it. Pointing the Virtmed team in that direction was my job. I pushed them to go out and spend a lot of time with customers and build the beta product, which allowed them to answer these questions. The results that they got back put them squarely in the 8–9 range.

That's what the dog food test is all about—it doesn't predict success, it evaluates the *possibility* of success. You still have to go out and do the hard work to prove it true.

Gathering and Using Customer Feedback

Obviously the big focus at this early stage should be on developing a solid product. Forget about the big investors, forget about the fancy management. Right now you should be worrying about whether or not the dogs—your potential customers—actually like the dog food.

Say you've scored high on the dog food test. But how do you know for certain whether or not customers will actually like your idea? Ask them. And do it before you sink resources into building the prototype.

I learned this from one of my entrepreneurs, Greg Gianforte of RightNow Technologies. Greg writes up the product or service idea on a one-page spec sheet and e-mails or faxes that sheet to ten to twenty potential customers and industry bigwigs. He calls and asks what they thought of his idea, then incorporates that feedback into the product or service. Then he modifies the one-page spec and sends it to even *more* prospects. By that point it's time to build a prototype and start beta-testing or selling. Then it can be modified based on feedback from customers who are actually using it.

These last few steps are the most important, and it's a process companies rely on repeatedly. I've combined them into four important steps, and I advise companies to waltz through them again and again and again.

The Four-Step Dance

---

1. **Call** prospects and gather feedback on the model.
2. **Modify** the model based on their feedback.
3. **Build** the model.
4. **Sell** it.

Then start the whole thing all over again. You can't do it too often, and it doesn't apply just to potential products. Say you already have an existing product. You can use the four-step dance right away to get more out of it. You'd modify the process slightly by talking to *existing* customers about their satisfaction level with the product, how they use it, and how it saves them money or time.

When you start talking to customers, you will often find that they're using your products in ways you never dreamed of, ways never mentioned in your sales literature or marketing brochures. When you find out about one of these new applications for your product, go back and test it as you did earlier, using the four-step dance. If you have a very positive product response (90–100 percent), you just found yourself the company's next product or application. The whole thing is a little bit like an Easter egg hunt: you've got to keep hunting around until you find the golden egg.

I'll tell you about a company that mastered the four-step dance. It's called Vellis, and I met the founder through Michael Begun of Coates & Meyer, an Australian venture capitalist who is a friend of mine. Vellis's founder had licensed the world's largest automotive repair database, which stores information on repairing parts and systems for all international trucks and cars.

About two years ago he and his team came to the ranch from Australia and talked about building a Web-based interactive site for companies, like Mercury Motors, Ford, or Navistar, that need to train employees and vendors on fixing the vehicle and its subsystems.

During the few days they spent in Montana, I pushed them on the four-step dance. I told them to get their butts over here and work with their potential customers—the big auto manufacturers in the States.

Six months later they did. The team moved to Chicago and started working three or four big accounts, including Navistar and Mercury Marine. They showed product demos on the PC and won a few trial accounts. Then they came back out to the ranch.

When the Vellis team first came out to see me, they were Dreamers. Pretty advanced Dreamers, since they had actually gone out and done some preliminary work with customers. But they still couldn't answer any questions about how customers would use their product. They didn't even know if the prospects used computers in their training process.

But the second time the team visited the ranch they had it down cold. They knew exactly what customers wanted and were in the process of adapting the product (adding features and functions) based on that feedback. They had even more customer commitments once those changes were made. At one point in his presentation the founder turned to the whiteboard and drew a chart of Navistar's development chain and at what points training occurs. That's a man who's done his homework.

In having such large prospects, Vellis had a real advantage. Each account was like having fifty accounts once we factored in the number of vendors and suppliers. Mercury Marine alone planned to put over three thousand people on the system. When Vellis got a suggestion to make a change, that change wasn't likely to be client-specific, since so many other companies would be using the system.

Vellis has secured several million dollars in financing, but they're not resting on their laurels. They still consider themselves to be in product trials (although now they get paid for the product). It's an ongoing process, and it's in Vellis' interest to dance fast. After all, the more Mercury Marine uses the product, the more time and money they have vested in Vellis's system. That makes it tougher for competitors to get in.

FINDING THE BEST PROSPECTS

The bottomless well you have to tap is customers and potential customers. To get the most out of them, you have to learn the right way to talk and the right way to listen. The first step is to find potential customers who are worth talking to.

What I advise is to make a list of the top target markets for your product. Then list the top three companies in each. And I really mean the top three—the market leaders—not the five you think will take your calls.

The next step is getting in to see them. Start flipping through your address book and calling friends. If you come up with *anybody* at that company, then you're ahead of the game. Call or e-mail them a very brief (three-paragraph) message outlining who you are and what you want. Something as simple as, "My name is Steve Smith and I met you at that cycling benefit last year. I'm working on developing a new travel services database; who would I talk to there about that?" Then you just e-mail and call your way up the food chain, referring to the insider who steered you toward each contact.

All right, say you didn't come up with a name. You just

have to work a little harder. First, you have to figure out who will help you the most—VP of business development, sales, marketing, or engineering. Then you have to get the name.

Be a detective. Check out the company Web site first, in the "About Us" section. Search business and trade publication sites to see if recent articles list helpful contact names. Large companies used to dealing with the press and public relations often happily give out the information over the phone; just call the receptionist and say matter-of-factly, "Hello, could you please give me the name of the marketing director?" It's worth a try.

> **TIP**
>
> **Don't underestimate the importance of assistants. They will move mountains, as long as you're respectful, polite, and not too obnoxious. An assistant who's your friend will tell you the best time to call, transfer you to someone else who can help, and sometimes argue your case to the boss. Be charming but tenacious.**

Tightly held companies can be more problematic. In that case I advise taking the direct approach. Call and tell the receptionist that you need to talk to the VP of engineering, but you don't have that person's name. This will work about half the time. Sometimes you get bumped to an assistant. Use the direct approach again. Explain briefly who you are and that your contact is not expecting your call. Sometimes I just say, "Your boss probably isn't the right person for me, I'm calling about X." Half the time the assistant says, "You're right, he's not. But let me transfer you to Bob Z, who will be able to help you." Politely ask if the assistant can possibly arrange just a few minutes on the phone. Be prepared to fax or e-mail a one-page description of what you're about.

When you finally get someone on the phone, but it's not the right person, don't hang up. He or she can probably help you find the right person. Ask for another name and if your contact wouldn't mind giving a heads-up that you're going to call.

GETTING THE MOST OUT OF THE MEETINGS

It might take a while, but eventually someone will agree to a brief meeting to talk about your idea. Arrive five minutes early, and bring a small thank-you gift (candy, flowers) for the assistant or receptionist who helped you get there. Leave the Powerpoint show at home; just bring one or two slides or a paper flipbook (low glitch potential). If you do bring slides, either bring the equipment to show them or check with an assistant beforehand to make sure what you need is on-site.

The key is to leverage your team's strength. If you're packed with engineers, build a prototype to demonstrate. If you've got lots of marketing people, put together a flip-book with charts or a slide presentation. Talk about what you're *going* to build, and know when it will be 80 percent done and ready for beta-testing.

If you've got a product, or just a demo, take it to the meeting. But be open-minded about it—don't tell the client how it will work for them, let them tell *you*. A good way to start off is to say something like "This is what we have; does it interest you?" Be short and sweet, straightforward and honest. Let them talk; sometimes the best ideas pop up at the ends of meetings. Your main job at the meeting isn't to dazzle them with a presentation. It's to listen. A good meeting is where you talk 20 percent of the time and the customer talks 80 percent.

Brainstorming like this is how Ascend came up with one

of its most successful products. We went to one customer with an idea that we hadn't built yet. As we talked about our technology, the client started thinking out loud about problems he was having. I sat there and listened and suddenly realized that I had my new product. We could solve his problems by changing our idea slightly to build exactly what he needed. And that's what we did, after bouncing the idea off a few other clients. That product—a dial-up box for dial-up Internet connections—is what helped build my company.

If you don't have a product, be prepared to offer a timeline on delivery dates—after all, the prospect might want to place an order. If that's the case, then ask for a conditional purchase order (contingent on a certain delivery date or specific product features). This shows your board and engineers that a real live company is interested. Then, before you leave the meeting, create a reason to come back. Say something like "In four to five weeks we will have another milestone." Offer to check back and update the contact.

After the meeting, send follow-up e-mail immediately to the assistant and the contact. Thank them for their time and sum up what you learned from the meeting. If you've agreed to follow through on some action items, list what the next steps and dates are for those.

Last, don't forget the four-step dance. When you find out why the customer likes your product, go back and incorporate the feedback. (Don't forget to change the presentation to reflect the new features and positioning.)

CLOSING THE SALE

Okay, you've finally found someone who's interested in buying your product, presumably from the hundred or more industry people you polled. Now you should ask a few (from different industries and market segments) to join your beta-test program. One tester just doesn't give you enough data points. Two don't tell you if there's a pattern to the information. Three will give you enough information to draw conclusions. If you have to give a very deep price discount, do it. I prefer to see a useful, "name brand" account sign on, and that type of client will likely want something in return. Offer a special discount.

> **TIP**
>
> If you're giving away product to testers, and the discount is 100 percent (which is fine for a really good beta account), invoice anyway. No matter what the price, invoicing establishes a clear paper trail and makes your company look more professional.

After customers have tested the product—surprise!—they want to buy it. But first they need a few dozen changes. You should sit down with them and go over their "wish list." Some changes will be easy to make, so go ahead and make them. Others will require more research. Narrow that list down until the changes agree with your customer research and you can make them within a short time—a few days, say, or weeks.

Anything more complicated than that demands more research. You don't want to get stuck in making substantial changes for a customer unless the changes improve the product for all customers. Before you agree to anything major, do some research to see if that's something all customers would buy or if you'd just be creating a custom product. Remember that any sizable feature change needs to go through the four-step dance (call, modify, build, sell).

I always tell my start-ups to get a firm purchase order.

When you do product planning, a lot of people want a lot of features, but what separates the men from the boys is their willingness to open the wallet. That commitment means they've gone to quite a bit of trouble—they've passed paper through the company for signatures. Basically, they're on board.

When Customer Feedback Is Too Much

Well, we've talked quite a lot about shaping new ideas using customer feedback. It might sound as though customers are a fountain of wisdom for idea *generation*. Do customers ever come up with that killer application, that gotta-have service? No, I don't think so, I've never seen it happen. The problem is that lots of companies expect them to. That leads companies into a couple of product-planning traps.

If an idea is *too* unformed before you start parading it in front of customers, you're expecting too much of them. You're looking for the customer to point the right direction. That's just too overwhelming for them. Anyway, that's *your* job—dreaming up products that you have the expertise to build.

Another risk is that a company can get too caught up in chasing one particular customer for feedback. Before you realize what's happening, you're slavishly following that single account until you get worked into some narrow niche trying to keep that customer happy. Then you find out there's no other customer for that product you've just created.

You can tell you're heading down that path if you jump on a customer suggestion without doing any due diligence (talking to other accounts). If you're starting to add more than just a few small functions or features to a product or service, do the homework first and gather a few more data points to make sure the idea has wings.

Even if you're careful to listen to all of your customers, you run the risk of listening too well. If you begin to aim all of the company resources toward satisfying customers and developing new products and features they ask for, you may never look around for new opportunities. It's a problem big enough to have had at least one book written about it, and I won't go into much detail here since few start-ups face this problem early on. Suffice it to say that sometimes you have to break the rules, introduce a little chaos into the system. Send some people over in the corner and let them follow a hunch. Otherwise you get entrenched, and another start-up comes along and knocks you out of your market.

SAME OLD MONKEYS, NEW TREES

Product planning should be internally inspired and externally verified. But even there, start-ups make mistakes. It's a careful balance between focusing internally and listening to customers, and it's easy to lean too far in one direction or another.

When companies get too internally focused, the managers run around chasing their own tails like a bunch of screaming monkeys. The manager who screeches the loudest attracts the most attention, surrounding him- or herself with committees of followers. They all run around, "reprioritizing" ideas and chattering about what to do next.

Maybe the lead manager's idea is good, but the followers are so busy listening that they forget to do the four-step dance. The idea never gets tested. The company monkeys are too busy chasing each other around to talk to customers.

When a manager starts screeching about a new product idea, everyone turns and starts to follow. By this time the organization has institutionalized the monkey frenzy, and product planning ends up being driven by lots of screeching and chest beating.

The monkeys are well on their way to creating a jungle bureaucracy. Management is too distracted by the whims of the dominant monkeys instead of doing customer-focused product planning.

WHEELBARROW

Over on the other side of the jungle, start-ups are making another classic product development mistake. The marketing team is mowing down trees for the paper to print a zillion research and marketing documents, stuff that's also slapped up on the Web site.

What happens is that the company is trying to plan carefully—*too* carefully. Marketing writes a fifty-page product-planning document to send to engineering. Engineering creates a sixty-page response, full of scientific jargon. Everything circulates through the company for signature after signature. Pretty soon management is bogged down in bureaucracy, lugging wheelbarrows full of paperwork through the company. Lots of documentation, but no real work.

In the meantime, the market shifts and the company misses the window of opportunity. And the whole process starts over again.

SPINNING TOP

Up in the tops of the trees, the CEO of another start-up is spinning around wildly, reacting to input from the various company departments and employees. The CEO has so many interruptions that it's impossible to focus on a single development path.

This CEO is bombarded from all angles and just starts spinning around in whichever direction the last input came from. Whoever hits the CEO last is the one who drives the direction du jour. There's no methodology, no quantifying system, to discern a good idea from junk.

THE DEATH SPIRAL

I worked with one company that actually went through all three of these stages during their start-up. They had a good product—a fax infrastructure that they planned to sell to Internet service providers and phone carriers, who would offer the fax service to their customers. But then they slipped into the "monkeys" phase of product development. Someone in the company started screeching about offering the product directly to the end user, so this start-up decided to sell to both the consumer and ISP/phone carriers.

That tactic pulled the company in two different directions. It burned through its start-up capital. Then, in self-defense, it headed into "wheelbarrow" mode. Because the dual-customer approach hadn't worked, the CEO wanted to document and plan every single product-planning move. There were mountains of documents flying back and forth, and every department had to sign off on everything.

Meanwhile the money ran out and the first CEO left. An interim CEO stepped in and was in turn replaced by a third. That CEO got stuck in "spinning top" mode, as he ramped up the company learning curve. The company finally dumped the consumer end of things and raised about $15 million in two rounds of funding. Even with $7 million in sales, it's a sad story. Because the company wasted time, distracted by poor product planning and even poorer execution, it never realized its promise of offering many different services. Now, potential customers want more than just a fax service, and competitors who offer more have entered the field. This company will be lucky if it gets acquired for the total investment.

The key to good product planning is forming hypotheses about the potential target markets. The next chapter outlines the Sunflower Model, which shows how to establish your company's core competency and then spin out idea after idea. The key is to then go out and test those ideas by talking to customers.

Don't get hung up on any one possibility, because customers might tell you that there's no demand. Instead of focusing on your idea of the target market, be open to opportunity, to the possibility that the customer (or other influences) might take your product or service in unexpected directions. Serendipity can lead down some profitable roads.

EXERCISE:
Test Your Product Development Process

1. Write down a list of top ten candidates for your product by industry. For example, two from the financial arena, two from the service sector, and so on.

2. Set up meetings with these candidates. For each candidate, write down why they liked or didn't like your idea.

3. Incorporate what you learned from 1 and 2 into your presentation and executive summary.

4. Take the new presentation/summary out to ten additional candidates. Repeat steps 1–3.

5. Build Rob's dog food test for your product. Focus on

   a. idea—write it down in one sentence (a short one).

   b. customers—list the top ten representing several industries ("top" refers to both size and prestige).

   c. money—write down precisely how you make money (how much and under what conditions).

   d. application—identify the business problem you are solving. Write it down and be concise; no hand waving!

   e. uniqueness—how do the customers use this product or service today? Who else does the same thing you're doing today? Who else will be doing it tomorrow? Build a list.

   f. value—write down precisely how and how much your customer makes or saves with respect to using your product or service.

   g. barriers—what, precisely, is your barrier to entry? Write it down.

   h. strategy—write down how your product impacts your customers' bottom line.

   i. competition—build a matrix. On one axis list all the companies that do what you do. On the other axis list the features and grade of what they offer. Compare the scores, and remember that a tie goes to your competition. To beat them, you have to be miles ahead.

   j. scalability—build a list of where your product might grow within your client company (and its supply chain and sales channels) or among client users.

▶ You are welcome to refer to the Entrepreneur America Web site for additional information (www.entrepreneur-america.com).

The Sunflower Model

ROAD MAP: ▶ You're on track to develop an exciting product that customers love. But what's the follow-up going to be? It's time to establish your company's core competency. How can you develop a cohesive family of products without knowing your company's strongest assets and abilities?

In early 1998 I invited Evan Thornley to Entrepreneur America. Evan, a former *Reader's Digest* magazine executive, was running a small Web "portal" company called LookSmart. He was smart, experienced, and ambitious, but LookSmart's search engine Web site wasn't really on the map. In fact, Reader's Digest, which had been funding LookSmart, had lost confidence and pulled the plug. Dwarfed by companies like Yahoo!, Evan's company was pulling in revenues of only $120,000 a year in advertising. The start-up was just barely staying alive.

"You're focusing on the wrong product," I told Evan bluntly. "You're never going to make a significant dent in the portal business." By thinking of himself as just a portal company, LookSmart had to worry about how to compete with Yahoo!, Lycos, Excite!, and half a dozen others. "You can't win at that game because others have already won," I told him. I'm not even sure what they've won. Even Yahoo! hasn't proven itself successfully in terms of profitable revenue.

We sat across the conference room table from each

other as I talked. "What is your core competency?" I asked. "What do you do better than anyone else?"

At Entrepreneur America we talk about core competencies as a way of understanding where product ideas come from and what the future products and markets should be. Establishing that drives the company in a positive direction, aiming top-level products and services at responsive markets.

Evan thought about my question for a minute. Finally he answered, "Well, we're different because we really know content." It turned out that LookSmart was great at finding and collecting links to other Web sites. They had lots of smart, highly educated editors (the lead medical editor, for example, is an M.D.) who spent their days searching for good sites to use as links. As a result, the categories—like "Travel"—on LookSmart's Web site had more (and better) links than almost anybody else.

As Evan and I worked, we began developing LookSmart's Sunflower Model. Here's how it works: Each part of the flower represents a key element of the company. The **center** represents the company's core competency. The **petals** represent the various products and markets. The **stem** represents the underlying assumptions that the business model is based on.

As we developed the model, it pointed the company in the new direction of distributing the content to other businesses. Before long, LookSmart entered into a nonexclusive licensing deal with Microsoft, which pays the company to provide it with content for its Web site. By mid-1999 LookSmart had made several similar deals and lined up $60 million in venture financing, based on a valuation of $370 million.

The lesson here is that entrepreneurs can easily get so wrapped up in the product that it's hard for them to think in terms of core technology or competencies. But anyone who wants to build a top company has to.

Planting the Sunflower

At Ascend, working with the Sunflower Model taught me that successful companies don't just roll out new products; they crack whole new markets. I learned that the role of a CEO is to create chaos of a specific kind—the creative chaos that leads to new ideas or leverages your old core competency into a new market. The CEO has many jobs, but one of the most important is to meet this goal: *Create one new market/application and associated product each year.* A slightly different product for the same market doesn't cut it. But modifying your product to sell to an entirely new market does. Opening up new markets creates strong new revenue streams. The kind of thinking and exploration that this goal creates will keep your company fresh and alive. Creating the new products will keep it healthy.

Finding other products or services is something entrepreneurs often find difficult. They feel so overwhelmed trying to get just one product out the door that they can't imagine sparing the energy to dream up others. But you have to start thinking about these things early, because it takes lots of time to plan and implement new products. That's where the Sunflower Model comes in.

SUNFLOWER CENTER:
ANALYZING CORE COMPETENCIES

When I do the Sunflower Model exercise with my start-ups, I draw the outline of a simple sunflower on the whiteboard and then ask the entrepreneur to step up and write the company's core competency in the center of the picture. What usually happens is that he or she grabs the marker and writes down the product name. Wrong answer.

I want to know what it is that you do well, better than anyone else. Your core competency is the underlying principle of the business model. For example, look at a really strong one—Priceline.com. The way the site sells is that customers type in the price they want to pay for an item (airline tickets, for example). Priceline.com matches them with someone willing to sell at that price and takes a percentage of the transaction. So what's the core competency—airline tickets? Sales? Nope—it's "The buyer names the price for an item." That's the key thing this company does.

> ▼ **TIP**
>
> **Look for the signs of a really strong core competency. Is it one brief (jargon-free) phrase? Does it turn a traditional paradigm on its ear?**

Here's another example: Honda. Is it a motorcycle company? A lawn mower company? An automotive company? No, its core competency is *building great combustion engines*. The lawn mower, the car and motorcycles, those are petals—the products and markets that grow from the core.

SUNFLOWER PETALS: PRODUCTS AND MARKETS

Once you've figured out your core technology, it's time to focus on how many ways you can leverage it. This is the moneymaker of the whole exercise, because you see not just the one or two applications that you originally planned for your product or service, you discover many other options. Instead of having a sunflower with a single petal, you grow a strong one in full bloom.

Take a look at the center of your sunflower. How many markets and applications can you think of where it might fit? Think about Honda again. The core competency is "building great combustion engines." I could draw petals marked "automotive," "marine," "aerospace," "farm and industrial" . . . the list could go on and on. I have a lot of experience creating petals—I end up doing them for a lot of my Technoid Wanna-bes. They usually don't know much about their markets and applications, so they really struggle with this part of the exercise. Then I send them out to do the footwork and test our theories.

Rank the petals according to their potential. Just analyze the fit—is your core competency a big or a small component of the petal? In the Honda example, engines are one of the key components to building a successful car. That means Honda is deeply embedded in that application, and that petal wins a high ranking. "Aerospace" ranks a little lower. Engines are a core component in building small aircraft, but even though the space shuttle might use a small combustion engine or two for running backup equipment, they're not a key component. Automotive earns a top rating, aerospace scores lower.

A few petals might drop off completely during this process. That's okay; the ones that are left will be stronger for it. Now that only the most promising are left, it's time to rank the markets. Are they strong, growing industries? Is there room for a new entry? Give them number values from 1 to 5, 1 being the best.

Once you've done that, look carefully at the top markets and pick your targets. Who are the industry leaders? Can you partner with them? Now you're beginning to find your new product or application. Talk to those accounts. Using their feedback, reevaluate whether or not that petal still belongs on the sunflower.

SUNFLOWER STEM: UNDERLYING ASSUMPTIONS

The stem, which supports the flower, stands for the underlying assumptions on which a business is based. Every business has them, whether they dig deep enough to expose them or not. Let's look at the example of Priceline.com, an Internet company that matches buyers with sellers on its Web site. For their business model to succeed, the company has to form certain hypotheses about the customer and market.

For example, "There are goods that sellers, after a certain time, cannot sell. Like airline tickets and food." This is a true underlying assumption—Priceline.com's model depends on it. Another might be "Sellers will sell the same goods at different times for different prices." If either of these truths were to change, Priceline.com would tank.

The common mistake is to take either too lofty of a view or focus in too tightly. I call the former the "hundred-thousand-foot view," when the assumptions end up being too trivial to actually relate to the specific company. For example, Priceline.com might write down on their stem something like "Consumers want to buy things online" or "Consumers will continue to feel comfortable with e-commerce" or other assumptions about bandwidth and the growth of Internet use. These underlying assumptions could apply to thousands of e-commerce businesses. It doesn't help your specific company to talk about them.

The flip side is to take too microscopic a focus, which I call the "ten-foot view." An example for Priceline.com would be, "Single white female consumers buy airline tickets more often than orangutans." Who cares? This focus is so tight, it assumes that Priceline.com's business model works only for single white females buying airline tickets.

Without the stem, there's no flower. So if these hypotheses turn out to be wrong, you're out of business, building something for a market that just doesn't exist. That's why it's important to identify the assumptions, then test them to make sure they'll keep supporting your company the way a stem holds a flower.

Making Your Sunflower Bloom

So, now it's time to put all this into practice. One thing I tell my Wanna-bes is to test the petals. You can't decide whether or not a petal will work while you're sitting in an office somewhere. Get out and talk to the leading potential accounts within that petal's market.

Once you've decided to go for it and develop the product or application, don't try to get it perfect the first time. Get it working 90 percent, which means that most of the features are in place, and those work 100 percent. Then get out the door quickly to test it with potential customers. Listen to what they have to say, then bring the product back in and fine-tune it before rolling it out to everyone. You'll have some misses; some stuff won't pan out and you'll have to drop it. But that's how a company learns, and it's better to take a chance on some wild idea than to lock the company into existing products, with nothing new in the pipeline. Pretty soon some small start-up is going to pop up and eat your lunch.

To keep employees in the sunflower mind-set, work on creating a learning culture. Listen to people who come up with the crazy "let's see if this flies" ideas. (You should also be coming up with your own.) Establish ways to reward speed when developing new products. That way the company institutionalizes this creative, chaotic way of thinking. You'll be the one surprising your competitors instead of the other way around.

One way to maintain the existing business while also pursuing new petals is to separate the petal development team from the rest of the company. I did this with Ascend, but it's definitely not easy. When I announced that we were adding dial-up Internet access to Ascend's videoconferencing capabilities, 20 percent of my sales and marketing people quit. They thought I was crazy to change course.

They also thought it would mean more work for them (they were right on that one).

The truth is that people are resistant to change, and if you want to introduce a new petal at your company, the way to do it is to communicate, communicate, communicate. As an instrument of change, you're going to get a ton of resistance. Resist the resistance; you have to be resolute if you're going to push it through.

> **TIP**
>
> Introducing change is easier if you sign up key supporters in advance so that you're not fighting the battle of one against the many. Get a great internal manager, an entrepreneurial type, to lead the effort. Keep the change team small, highly focused, and working with firm deadlines so that the company can see that you're serious about it.

Sometimes I see companies really struggling with creating their sunflower. Most often they're simply unable to identify their core competency. Instead they keep falling back on the existing product, insisting that's the core asset. That problem can be solved, fortunately, by following the exercise at the end of this chapter.

What's worse are the companies that don't even have a core competency—what they're "best" at is something a million other companies are also best at. This is a very slippery rock on which to build a company. This is where I typically see a lot of Quickie Wanna-bes. If they're lucky, they're first to market. Enough lead time over competitors might buy some time to grow a company. A good example of this is eBay—there's nothing really mystical or magical about on-line auctions. But they got there first. By the time competitors popped up, eBay had developed some core competencies: massive on-line auction skills and clever tracking of people when they're on-line. The company had a pile of expertise plus a brand.

SUNFLOWERS FOR START-UPS

It's never too early to create the Sunflower Model. At Entrepreneur America we build Sunflowers for companies that haven't even settled on their products yet. One of my Guts and Brains Wanna-bes that ended up developing a particularly good one is called RightNow Technologies. They're based in Bozeman, Montana, and founder Greg Gianforte drove out to meet me almost two years ago.

Greg's idea was to build Web software for customer relation management on RightNow's customers' e-service sites. The software would help customers with incoming e-mail questions (from *their* customers) and the outgoing answers. Basically RightNow wanted to be the infrastructure that answers when customers e-mail Ben & Jerry's with a question about where the nuts come from for Rainforest Crunch ice cream.

About a year later, once several companies were actually using the RightNow software, Greg and I sat down to work up his sunflower. We wanted to explore the full story of where his company could go with the software he had developed.

First we looked at the core. His core competency was clearly the software, which had some impressive capabilities. Also, the founders had strong experience with building an in-house telesales force at their previous jobs. We wrote, "Customer relation management software sold through telesales" in the middle, as shown in slide 1. There were a few other angles RightNow was working on to gain an edge over its competitors. First, Greg's software was easy to install and use—customers could be up and running with the thing within forty-eight hours. Also, Greg planned to move in the direction of being an application software provider, meaning that he'd host the software himself (basically clients would outsource the whole customer e-mail support function to him). We didn't write those down, though, because although they were key strengths, they were peripheral to the real core competency.

RIGHTNOW'S SUNFLOWER CORE

Slide 1

Clearly this core competency had legs. As we thought about the petals, we started thinking of that word *customer*. It could mean a lot of different things, not just ice-cream eaters e-mailing questions to Ben & Jerry. It could be investors asking Fidelity about how to calculate earnings ratios. It could be vendors asking manufacturers about materials. Or internal salespeople asking engineers about product specs. Wherever any corporation had a question/answer requirement, RightNow could be in the loop.

We started drawing petals like crazy (see slide 2). The first one was "customer support," which is what they were already doing. The second was a subset of customer support, "tech support" (questions about how to use the product). Then "marketing support," which would be questions about product features, availability, price, and so on. We drew a petal for "internal help desks," which answer employee questions about using company computer systems. Where else do companies have questions? How about "human resource management"—questions about employment and benefits?

Greg and I put ourselves in the shoes of his clients, and we brainstormed on all of the possible places where a "customer" (internal or external) might have a question. For each of those, we made a petal.

Then we ranked them. For the "sales support" petal, for example, e-mail management (which is what RightNow does) is only one key component. The other angle to evaluate the petal is in light of competition, and sales support has a ton of powerful, well-established companies already vying for market share. We gave this petal a low ranking, around a 4 on a scale of 1 to 10.

"Marketing support" ranked much higher. It depends heavily on the e-mail management aspect, plus there aren't many competitors in the field. That one scored 8-9, a potential area to begin studying.

Finally, we looked at the stem—the underlying assumptions on which Greg was basing his entire business model, as shown in slide 3. First we assumed that companies would continue to process lots of questions—externally and internally. Well, that's a no-brainer. Greg's model made two more assumptions: that companies would continue to outsource the entire "customer service e-mail" function and the continued efficacy of the telesales model.

Now, these assumptions probably won't make or break the company. If it turns out that clients would rather have a richer, more functional solution and they're willing to wait five months for the installation, Greg could probably swim upstream in that direction. Or if they aren't willing to

RIGHTNOW'S SUNFLOWER WITH PETALS

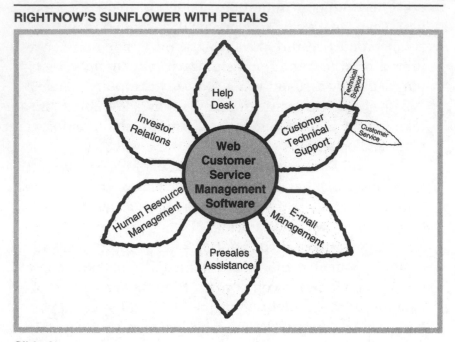

Slide 2

outsource their back-end customer support, he could focus on just selling the software instead of the service. If telesales turned out to be an ineffective sales model, he could retool and ramp up another approach. But in order to make any of these leaps, Greg had to first recognize what he was doing. He had to clearly identify what assumptions he was making, so if they turned out to be false, he could make the change.

SUNFLOWERS FOR MATURE COMPANIES: THE FLYWHEEL PHENOMENON

It's not hard to see how the Sunflower Model can also work at big, established companies. Instead of focusing on extending existing product lines in the same old directions, the Sunflower Model pushes big companies to think about how to take their core competencies and push them into new markets and applications.

There's only one problem. At established companies it's much harder to get authorization and resources to proceed with exploring new petals. Why? The Flywheel Phenomenon.

Once a company is entrenched, it's like a flywheel moving around and around. All of its momentum (resources) gets focused on existing customers. What do they want? What new features are they asking for? New product development is prioritized according to the largest customers offering the biggest opportunity for more sales.

What never really comes out of a flywheel is brand-new market generation. Since it hasn't grown out of customer

RIGHTNOW'S COMPLETE SUNFLOWER

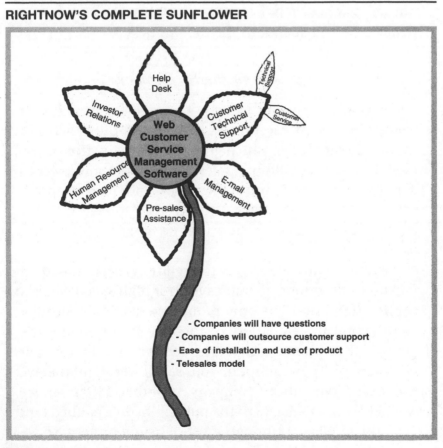

Slide 3

demand, you can't really demonstrate who will buy it, how long it will take to develop it, or how big the new market will be. Without hard stats, flywheel companies will rarely put the resources into launching a weird little idea that grew out of a sunflower petal.

A good CEO won't let that happen, because he or she pushes for change and forms some kind of skunk works independent of main company body. If there's no institutionalized way to pursue new ideas, breaking free of the Flywheel Phenomenon to follow an independent idea is hard to do, but the sunflower can help. By developing that model, you can demonstrate how strongly your little idea relates to the company's core competency. Management can see that you're not coming up with some far-out idea that completely abandons the old business.

Harvesting the Sunflower

The reason I apply this exercise to my start-ups is that it worked for me. At Ascend we really reaped the benefits of the Sunflower Model, which grew the company from a few million to several billion in sales. We used the sunflower to identify new markets and applications—not just new products—that we could attack. It turned out to be a valuable lesson that I learned about a year after starting the company.

At that point Ascend was struggling. We had raised $3 million in VC money to build a box for high-speed data access to ISDN lines. The only problem was that about nine months and three-quarters of the way into building the product, the phone companies decided not to roll out ISDN service. The assumption (our sunflower stem) that we had based our whole company on—that ISDN service would become available to the public—hadn't panned out. Ascend had burned through most of its investment, so we were broke, with an unfinished product and no market.

Desperate to find more money, I decided to give new investors the same pitch—our ISDN box business. But I added a small caveat: "The ISDN service might roll out a little slower than we expect." On the strength of that story, we managed to raise another $3 million.

A month or two later, I finally found two customers. Boeing and MIT bought a box each—they ran their own internal phone companies and so had their own ISDN lines. But Ascend was clearly in trouble. I sat down with my co-founders and started brainstorming. How could we salvage the company?

One of my co-founders had the idea of building "bandwidth on demand" for data transfer over regular phone lines. The great thing about the "demand" part is that your computer (or fax machine or whatever is sending data) gets more bandwidth as it needs it. Basically it's like adding lanes to a busy highway as traffic increases and removing them as traffic wanes (except that we're talking about dial-up phone lines).

At the next board meeting it was time to come clean with the investors. "I have bad news and good news," I announced. "The bad news is that we've terminated the product." Silence. I started to sweat, wondering if I'd completely screwed up. Then the most recent investor, Dick Kramlich of New Enterprise Associates, started laughing. It was his first meeting, and already his money was on a dead product. Dick dug around for our business plan, pretended to read it, and then tossed it in the air in mock frustration. "I just love entrepreneurs," he said, smiling. "So, what's the good news?" His sense of humor and trust completely saved us.

The good news was that we had another product. Well, at least an idea. My team and I took to the whiteboard to explain the dynamic bandwidth concept, which at that point was just a figment of our desperate imaginations. We hadn't talked to any customers. The board listened patiently and ultimately decided to give us three months to prove the concept—both the technology and whether or not there was customer demand.

At that point we weren't thinking about core competencies or sunflower petals or anything but saving our butts. After the board meeting I divided my team of five engineers into two squads. One team of three pushed into new product development, the remaining two would finish the original ISDN product (hedging our bets).

Next we started brainstorming about what markets could use this dynamic bandwidth product, something that didn't need to be connected twenty-four hours each day. Several came to mind, including videoconferencing transmissions and connecting corporate offices to each other. We looked at videoconferencing first because it seemed like a good fit. It was used only a few hours each day, and the quality was driven by bandwidth. A company that wanted higher quality could pay higher prices for more bandwidth.

To explore the idea, we made a list of the top companies in the videoconferencing industry. A few days later we got fifteen minutes with the industry leader, John Tyson of Compression Labs, which controlled 55 percent market share. We walked in with our concept clearly described on a single sheet of paper.

John got excited because his videoconferencing units required high-speed connections, which he secured as leased lines from the phone company at $5,000–$7,000 per month. Customers had to rent two of those lines, in addition to buying Compression Labs' specialty videoconferencing equipment. Clearly our concept would save everyone money, and John could sell more video.

When someone like John gets excited, I always ask for an order. I asked him to buy two units. Then he wanted to know the price and delivery dates. I took a stab at the pricing, asking for $15,000 each. I knew we could build it for a profit, and it would still be a great bargain for John's customers, who spent that much in two months renting the expensive lines.

As for the delivery date, I had to admit that we hadn't actually built the product yet. I told John that his check—an endorsement from the largest video company in the country—would convince my board to let me build it. He agreed to write the check (and I agreed not to cash it until product delivery). For that, I'm eternally grateful to John.

I used that check to prove to our board that we had impressive customer interest, which bought us another nine months. And we pulled it off. We built the product and installed it with Compression Labs and other customers. Ascend grew from zero to $16 million in sales (and 50 percent of the high-speed videoconferencing transmission market) on the back of that product.

But we weren't a great networking company. We were just a big fish in a little pond. The board started making rumblings about selling Ascend, cashing out for a modest amount and calling it a day. By that point I had raised a total of $20 million in return for 75 percent of the stock, but we weren't giving our investors the ten-times (minimum) return they wanted. To do that, I knew that I needed to reinvent my company once again. We needed a bigger market and more revenue.

That's when I read an article in the *Harvard Business Review* about how Honda leverages its core competency of building engines. They think of themselves not as a car company or a lawn mower company, but as an *engine* company. I realized that if Ascend was going to survive and grow, we had to build and harvest our own sunflower.

I started thinking. What was Ascend's core competency? We had built a box for ISDN (aborted), then a box for videoconferencing transmission. So what were we: an ISDN company or a video-transmission company? I struggled with this problem for several days until I finally realized that we were neither. What Ascend knew how to do well was send data over dialed-up digital phone lines, working with any telephone switch in the world. Our core competency was "dial-up networking." Once I had that core, I sat down the following week with my co-founders and drew up the rest of Ascend's sunflower.

We came up with several new petals. It was 1991, and the Internet wasn't a big deal yet. But we felt that it would grow. The first petal we wrote down was "Internet access." That was a strong start; what else? We also thought of "remote access" for corporations connecting to small offices and "telecommuting" for people working at home who needed to connect to the main office.

Then we spent some time analyzing our stem, the underlying assumptions of our business. When we started Ascend, our underlying assumption was that ISDN was going to be available to consumers. Then the phone companies pulled the rug out from under us. I didn't want to make that mistake again.

With the dial-up business, our stem was based on several assumptions. First, phone carriers would continue to provide dial-up services. Second, those services would be properly priced. Third, there would be applications that would require high-speed dial-ups versus a fixed connection. If any one of those stem assumptions proved wrong, we'd be out of business, because we'd be building equipment for something that didn't exist or was priced some crazy way. There's only so much you can do to predict trends, but based on our knowledge of the industry, these looked like pretty solid assumptions.

We started putting the sunflower into action. Because the Internet access market had the strongest potential, we started there. We spent the next six weeks flying around the country, talking to about fifty Internet service providers. Basically we walked in and talked about our core competency of dial-up networking and our old product, the videoconferencing box. We asked each of them,

"Can you use it?" At PSINet, one of the largest ISPs, we scored.

Marty, our contact at PSI, showed us his company's back room—a tangle of modems and snaking wires for his dial-up customers to connect to the Internet. The company was growing so quickly that it could barely keep up in adding modems and phone lines. Marty was excited about our video product because the one tidy box could replace his clutter of modems and wires.

He asked for a small box that could handle ninety-six dial-up connections. He actually sat down and designed what he wanted. Of course, when I saw his interest, how could I say no? "Sure, we can build that," I said. "Will you give us an order?"

Well, then he asked me the price. Good question, considering we hadn't even built the thing yet. When in doubt, I always move toward value pricing rather than cost pricing. I asked Marty how much it currently cost him for ninety-six dial-ups, and he told me $100,000. I priced our box at $55,000—I knew he would buy it at that price, and we could build it at a fat gross margin.

We built the box in six months. Marty was so happy with the product that he started telling others in the industry, and pretty soon we had a feeding frenzy. Ascend quickly grew from $16 million to $40 million, $150 million, $600 million, and then $1.3 billion; by the time it was acquired by Lucent in 1999, it was worth over $22 billion.

The growth was all based on executing our sunflower. We followed the same customer-led development process for the remote access and telecommuter markets. Before we realized it, Ascend had penetrated Cisco's domain—through the back door. We had 29 percent market share, and they were at 28 percent. We never set out to grab those markets through the front door, announcing that we had an Internet router to compete with Cisco. If we had, Cisco probably would have looked down and crushed us. Instead we worked the sunflower and crept in quietly, leveraging our existing strengths to build a backroom box that Cisco didn't even notice until it was too late.

By the time I left Ascend we were working four petals for sales of $150 million, growing rapidly to $600 million. We never dropped our main product, the videoconferencing device. But we added other products, every single one based on our core competency of "dial-up networking." Once I figured out that Sunflower Model, I was hooked, and it made my company.

A company that has mastered what's covered in these first three chapters—a strong team, solid product, and well-leveraged core competency—has the tools to get funded. The next step is to assemble them into the smartest package. The next chapter explains how to position the product or service for maximum effect, ensuring that it offers strong value, stands apart from competitors, and will grow the way investors like to see.

EXERCISE:
Planting Your Own Sunflower

---

1. Clear your head. You don't want to be distracted by worrying about your current product or service.

2. Think about what it is that you do really well, better than anyone else. The thing that makes "you" you. Write it down. For example, "We are good at making software and hardware that allows for dialing using AT&T switches." Notice that I didn't toss around any marketing brochure buzzword phrases. I also didn't talk about any products, just a *skill*.

3. Draw a circle. Inside it write the phrase. That's the center of your sunflower.

4. Draw some petals shooting out of the center. Write your existing idea/product/service on one of the petals. Then put it out of your mind completely.

5. Let your mind wander as you study the center of the sunflower. Write down any other application that could be built using the core as a major piece. Don't worry just yet about how feasible they are, just write them down on the petals around the core.

6. Now rank the petals using several factors:

a. How important is your core competency to the success of that petal (market or application)? Remember the Honda example—automotive ranks very high because engines are an integral part of its success. Aerospace ranks lower.

b. Look at the rate of growth of the petal's market. Fast-growth markets rate the highest.

c. Examine the competition. The less of it there is, the better.

d. Think about what new resources would have to be added for your company to dive into that market—time, money, employees, and other resources.

7. Focus on the highest-scoring petals. Those are your new markets, the ones where you have the highest probability of return.

8. Once the priorities are set, make a list of the top ten accounts for each of the high-ranking petals in the market.

9. Go out and speak with them. This can give you feedback on fine-tuning your rank of the markets.

10. Don't forget to revisit your sunflower each year.

▶ You are welcome to refer to the Entrepreneur America Web site for additional information (www.entrepreneur-america.com).

4 The Keys to the Gold Mine

ROAD MAP: ▶ With a strong core competency identified, it's time to finish your homework. The last step before talking to investors is developing powerful answers to key questions about your product's value, differentiation, scaling, and stickiness.

One morning I got a call from Alex Smith, founder of a start-up I had been working with. His company, Vellis, creates skills and knowledge distribution systems for companies in the automotive industry—basically they do business-to-business "e-learning," or on-line training. Alex had received a term sheet from Sterling Capital, a Chicago-based VC firm. The problem was that Sterling had stalled out. Although they had sent Vellis a funding proposal thirty days earlier, they decided to back out. Vellis was losing momentum.

Vellis had done its homework really well. They had a strong team that made it to the Guts and Brains Wanna-be stage. Customer trials had established that the dogs did, indeed, like the dog food. And the start-up's Sunflower Model showed several promising markets for the company products. But the next step in the process is to clearly demonstrate the product's value, differentiation, and growth potential of the investment opportunity (scaling and stickiness).

At this stage in Vellis's development, I needed to jump in, and fast. I had just joined Vellis and hadn't been involved in the Sterling deal. I wanted to turn the problem around and land a top-tier VC group for my start-up.

I checked the state of our "VC collateral," the executive summary and business presentation. They were pretty miserable, so I rewrote them, streamlining and punching them up. Then I e-mailed it to two VCs, one at New Enterprise Associates and another at Worldview.

A few days later I called up the first one, Mark Perry at New Enterprise Associates.

"Hi, Mark, it's Rob," I started off. "How much time do you have?"

"Not much," he said. "I'm in the car heading to a board meeting."

I worked quickly. "Did you get a chance to read the executive summary I sent on Vellis?"

"Yes, and here's what I want to know," he said. "Why do you like the deal?"

I figured I probably had about two minutes. Just enough time to highlight the four key elements that VCs look for: value, differentiation, scalability, and stickiness. If a company has those, they have the keys to the gold mine.

"Four key reasons," I answered. "First, they have a great value proposition. With Vellis, automotive companies can search on-line for training material, get refresher material for specific courses, peruse technical publications, and order parts. Vellis sells to three different budgets at each of its corporate customers: warranty, technical publications, and training. Each of these has budgets in the hundreds of millions of dollars. Vellis looks heroic because they save customers a huge chunk of change in each budget.

"Second, they've got major differentiation. Their proprietary database can't be replicated by competitors overnight, and they're constantly updating it through their customer relationships.

"Third, I love the scaling. Once customers start using

this training model, their resellers and supply chain partners are going to see it and want it. The whole thing will
spread like a virus. And fourth, the whole thing is really
sticky. Once Vellis trains twelve thousand technicians at
lower cost than traditional training, that's not something
easy for customers to replace."

Mark listened intently. Car noise hummed in the back-
ground as I wrapped up the pitch. Then he replied, "Okay,
Rob, I get it, and I want to talk to my partners." Later that
day I made the same pitch to James Wei of Worldview. Both
firms engaged in further discussion with Vellis.

It's true that not everybody can just pick up the phone
and talk to this caliber of VCs. But once I connect with
them, I use a "mental" executive summary to make a pitch
just as anyone else would.

Before the pitch, I write on a piece of paper "Value, Differentiation, Scaling, Stickiness" and a few key words after
each one. To be fundable, to get me or VCs interested in
your company, you've got to have some powerful answers
to these key questions.

Value: Setting the Gold Standard

The very first day I met with the team of RightNow Technologies, a Guts and Brains Wanna-be that makes the back-
end software for on-line customer service, I could see that
the company had a good product. But they weren't leveraging its true value to customers.

Because Greg wanted to spread RightNow to as many
users as quickly as possible, he had priced his product at
$15,000—well below the value to the customer. To com-
pound the problem, sales was regularly accepting $5,000–
$10,000 for a two-year license. Sure, they were gaining
market share. But they were giving the product away.

"Your pricing is way too low," I told them when they
visited Entrepreneur America. "We're going to raise the

price—double or triple it. In fact, eventually I want to get it up to around $100,000."

I could see that Greg Gianforte, the founder, was nervous. We talked about value pricing and calculating what your product is worth to a customer. Customer support employees cost about $100,000–$150,000 in salary, benefits, and equipment. How much money per month does RightNow's product save companies because fewer phone calls hit the customer support center?

Greg ran some value calculations and came up with an amazing figure: his customers were saving thousands to hundreds of thousands of dollars per month. Their customers were more satisfied and likely to buy more product. RightNow's price wasn't a factor in the decision to buy. In fact, for some customers the price was too low, making them wonder how the product could be any good if RightNow was "giving" it away. Greg realized that he could sell the product at *five to six times* his original price.

Because he was understandably concerned about making such a huge leap, I suggested that he call existing customers to ask if they'd be willing to spend more on RightNow's product. Each of the fifteen customers polled said "yes."

So RightNow raised prices to $29,000 for a two-year license, plus another $10,000 for various extra modules. We agreed on a sales process where discounts were strictly governed. Now Greg is on track to command a license price of more like $35,000–$50,000 per year instead of $5,000.

That kind of value proposition is quantifiable and clearly shows that it saves time or money. It's an easy sell to investors. The fuzzy wuzzy ones where people say, "Well, you'll do your job better," or, "We're better than our competitors," are impossible to quantify. The reason I don't like many Internet start-ups, particularly consumer models, is that they don't have a measurable value proposition.

Does your product or service fill a compelling need?

Lots of people have designed clever devices, things I call "widgies." You've seen or heard about these things, like a can opener that doubles as a fishing rod. These widgies might be fun to play with, but they aren't truly useful. They show how brainy and technically adept the inventor is, but they don't solve a compelling problem. I flat-out refuse to work with companies unless what they have saves customers lots of money.

GOOD VALUE PROPOSITIONS

RightNow's value can be clearly demonstrated with hard numbers. There are other aspects to establishing good value. I look for four elements:

1. A well-defined business opportunity.
2. A benefit or value that's measurable in dollars.
3. A favorable price/benefit ratio.
4. A short payback period.

First, a company has to meet an unfulfilled business need that corporations really care about and rely on. For example, RightNow's opportunity is driven by the fact that nearly half of the top-ranked Internet sites offer inferior *customer* service. Surveys indicate that 26 percent never respond to e-mail queries, and 29 percent take a day or more to reply. On-line customers service loads are triple that of telephone queries. Ben & Jerry's (one of RightNow's customers) receives five thousand queries a day. Clearly there is a well-defined business opportunity for RightNow to help their customers with e-service.

Second, the benefit of RightNow's product is demonstrable in hard dollars. It reduces cost in customer service head counts and dollars, increases customer loyalty and revenue, and builds a knowledge database by capturing the information from the e-mail question/answer exchange. In beta tests RightNow reduced Ben & Jerry's e-mail requests by two-thirds. It resolved 90 percent of the Web visitor issues without human intervention. Within thirty days, another client, Interland, reduced incoming phone calls from nineteen thousand per month to eight thousand. That saved the company $250,000 per month.

Third, the competition's product costs are three or four times more. So RightNow's price/benefit ratio is superior to that of its competitors.

Finally, RightNow's product pays for itself in months (or, in some cases, a fraction of a month). This is a very tantalizing payback period.

Here's another example. Silicon Spice is a Silicon Valley company founded by a young MIT graduate named Ian Eslick. While still in school, Ian came up with a design for a new "super"-semiconductor chip that could do lots of things at once, very quickly.

When I first met Ian, he hadn't actually built the chip. It costs several million dollars to build a semiconductor chip. In fact, to make ends meet, Ian was living in a Palo Alto garage that doubled as his lab. The company was pure Technoid. Ian was brilliant, and fortunately he was able to pick up business and marketing skills—a killer combination of brains and business sense.

I saw that Ian's idea had real potential in the telecommunications market. It offered the telcoms a great value proposition. A company like Cisco could buy just one of these chips, instead of a whole handful, to fit into their switch and network boxes that were crammed full of modem chips. Silicon Spice had a great value proposition for two main reasons. One, it reduced the number of chips a customer must buy and thus reduced the customers' cost. Two, it offered more capacity to add new software.

Applying my four steps of measuring value, Silicon Spice scores high. First, there's a well-defined business opportunity. The telcom industry is exploding, demanding more and more infrastructure equipment. Faster, cheaper, and better will always sell.

Second, there's a measurable dollar value. This new chip replaces a certain number of old components, so the savings can be calculated. Plus, intangible new functions can be added to the powerful chip.

Third, the price/benefit ratio is great compared with that of existing solutions. The chip is so much better that it outperforms anything else available. And fourth, Silicon Spice calculated the payback period to be about fifteen months.

I was so intrigued by Ian's invention that after we did our homework, I called up several Silicon Valley venture capitalists I know. After the normal due diligence, Ian and his young team landed $3.5 million in seed money for their new company. This makes it sound as if it's easy to get VC money. It's not. *One* reason Silicon Spice made it is because of the company's clearly demonstrable value proposition.

A good, hard value proposition is one that streamlines something that was previously very inefficient. Of course, this always boils down to money, in one way or another. It saves customers either dollars or time—each can be measured, and each directly affects the bottom line.

If your product or service saves customers money, dig for the dollar details. How much money, precisely, does the customer *save* per month or per year? How much does your product or service *cost* customers (measured against what they save)? What is the *payback* period (the amount of time before the customer earns back the initial investment in your product)?

If it saves time, quantify that also. How much time does the task take customers *before* they bought your product or service? How much time does it take them *afterward?* What is that time *worth* to them?

One key aspect of the value test is that the benefit be measurable by an objective calculation. To clearly address that, I developed a simple value matrix to apply to each product or service a company offers. A blank value matrix is shown in slide 1.

THE VALUE MATRIX

	Cost Before	Cost Now	Savings per Year	Payback
Money				
Time				

Slide 1

Here's an example: Virtual Ink produces a product called Mimio. It's hardware and software that allows a standard whiteboard in your room or conference room to become an input device into a PC. So any data on the whiteboard can be organized and stored on the PC. Anyone who has filled up a whiteboard with precious scribbles, diagrams, and graphs has run into classic problems: losing time making copies for attendees or losing data as you erase for more space (or others erase after the meeting). With Virtual Ink's product, there's no lost data or time. This is a productivity tool so obvious that everyone "gets it" right away. The detailed Virtual Ink value matrix is shown in slide 2.

Vellis is another example. This start-up produces Web-based automotive-training content and software that allows corporations involved with manufacturing autos or auto subsystems or large users of auto/trucks to facilitate

VIRTUAL INK VALUE MATRIX
Calculated at $50/hour per person (Average person $100,000 per year)

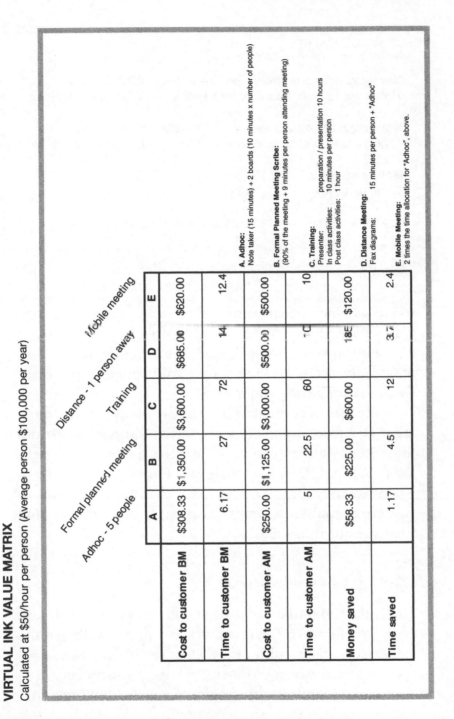

Slide 2

VELLIS VALUE MATRIX

> **Money cost to customers before Vellis:** $200 per day
> **Time cost to customers before Vellis:**   4 full days per class
>
> **Money cost to customers after Vellis:** $50 per day
> **Time cost to customers after Vellis:**   6-8 hours per class
>
> **Money savings:** $150 per day
> **Time savings:**   6-7 hours each day
>
> **Payback period:**   12 months based on one course

Slide 3

training of technicians and repair people. Before Vellis, several conditions prevailed: actual information and content was generally incomplete, and the training process involved physically moving people to instructors and classrooms, costing time and money.

With Vellis's content and Web software, training in the latest and most complete automotive content can be online. The instructors can be in Chicago, for example, teaching students in San Francisco and Boston who can learn whenever they like (as opposed to set times). The value to customers translates simply and easily into money and time, as shown in slide 3.

UGLY VALUE PROPOSITIONS

The truly bad value propositions are ones in which you can't get a fix on the finances. When you start filling in the value matrix boxes with phrases like "Huh?" or "Not sure," then you're in trouble. Time to rethink your strategy.

I started working with a company called Passlogix that had a product that stored computer passwords. Users would create a single password to log in to the software product, called VEGO, which would then retrieve all their other passwords and plug them in automatically whenever needed.

Originally Passlogix planned to aim this product at consumers, pricing it at $99. But we discovered that the value proposition on this was truly bad. What did customers do before VEGO? Write down their passwords on paper and tape that to the computer monitor. How do you determine what that effort costs a customer in money or time? You can't, it's too meaningless.

We revamped VEGO as a business-to-business application. Companies with help desks that spend time changing and handing out lost passwords could really use it. Some of those, we discovered as we talked with prospects, actually knew what that time cost them. Clearly VEGO could save a measurable amount of time (handing out lost passwords) and money (fewer help desk employees needed). Much better value proposition.

The worst example I can think of is an idea pitched to me on a global positioning system (GPS) beanie for the blind. Customers would buy this beanie, which had some kind of GPS antenna on top to pick up satellite signals. The beanie would talk to them, telling them where they were.

The value proposition is clearly bad. Sure, you can quantify how much the dog costs a customer (purchase and training price, lifetime food and vet bills, and so forth). And you can quantify how much the beanie costs. But the time spent "before" with the dog is actually an investment in a relationship. You can't quantify the attachment that blind people develop with their helpful companions. It's tough to form a lifelong emotional bond with a hat.

Even worse was the technology. GPS is accurate only to within a few yards—the difference between standing safely on the curb and being flattened by a truck. There's no value proposition in getting the customer killed!

I passed on this investment.

The Five Rules of Differentiation

Active Education, a small company founded by Mike Jossi, publishes technical training materials. The company reached a breakthrough when it began talking to its customers and discovered a profitable market supplying books and on-line materials to companies, like Microsoft, that outsource technical training to community colleges.

This was a step in the right direction, but it wasn't enough. Active Education had a good idea and a good product, but when they came to Entrepreneur America I could see that they hadn't differentiated themselves enough. Competitors could easily do what they were doing because they hadn't created sufficient barriers to entry. I knew that they would never raise significant money unless they could show how they really stood out, and I told them so.

"Tell me a little more about the product," I prodded the team. "What makes it really special?"

They talked about it a little bit, and then Mike started fishing around for details. He was getting creative, thinking of weird little aspects of the product that he didn't normally focus on.

"Well, customers really like that we can target a student's specific weakness," he said. It turned out that Active Education's on-line materials could shift to work with a specific student. For example, if a student stumbled with calculus, the software would automatically kick out more derivative equations to that student.

I got excited. That was very differentiating, but Mike and his team weren't mentioning this in their presentation to potential investors. They hadn't been thinking creatively, integrating this new and exciting viewpoint into their own perspective.

Lots of businesses have a differentiating idea, but they don't recognize how important it is. Or they just don't emphasize it enough. Their differentiating feature gets buried in all the other information and junk they throw out at investors and customers. At Entrepreneur America I often have to spend hours, or days, drawing this information out of people, slowly and painfully. So how do you know if your idea is unusual enough? You're all whooped up about your differentiation, but maybe you're just deluding yourself. The answer is to run your idea through my five rules of differentiation.

1. The Red Polka-Dotted Zebra: Do you have enough distinctive characteristics for customers to pick you out of the competitive crowd?

2. Ask the Right Questions: These will lead you to build a compelling product. The dogs will tear the package apart to get at the dog food.

3. The 10x or Better Rule: To get attention, your product must be either one-tenth the cost of your competitor's or ten times the performance. Anything less will not derail the incumbent company or technology.

4. The One Year or Longer Head Start: Your business model should be difficult enough to copy that it will take competitors a year or more to catch up.

5. The Improving with Age Rule: The differentiation should scale. This means that as you work with your business model (during your year head start) your differentiation keeps getting richer and deeper.

Some people think that just because they have an idea, they have a business. They don't realize that ideas are like belly buttons: everyone has one. But you have to be creative and look at issues with a fresh eye to be able to distinguish whether or not yours is truly different.

THE RED POLKA-DOTTED ZEBRA

Have you ever seen a herd of zebras on television (or actually out on the plains of Africa)? What is remarkable about their striped camouflage is that you can't tell one zebra from another. It's impossible to see which has a shorter tail, wider stripes, or darker mane. When the herd runs, it creates an optical illusion. It's harder for the lion to fixate on the youngest zebra, or oldest, or weakest. Instead she just picks off whichever zebra happens to be hanging around the outskirts. They all look the same to the hungry lion.

But what if I painted red polka dots on one of the zebras? Suddenly you'd have no trouble picking it out of the crowd.

Once when I was giving the "zebra" lecture at Cornell, a member of the audience raised his hand to speak. He turned out to be a biologist who had actually performed the experiment I was talking about. His research team had separated a few zebras from the herd and painted splashy red dots on them. They had then released them and gone off to bed, planning to study them over the next few days and find out how the herd reacted to the new paint jobs. But in the morning the red-dotted zebras were gone. Lions had quickly zeroed in on the dotted zebras and eaten them.

Most start-ups, like zebras, are herd animals. Their ideas edge into crowded industries. The products and services are variations on black and white with a few drab distinctions—a longer tail, a few more stripes. Their customers and the media can't see much difference among competitors. When faced with a herd of zebras to choose from, customers are going to behave like the lions. Either they'll just randomly choose a zebra from the edge of the pack, or they'll single out the zebra with the red dots.

ASK THE RIGHT QUESTIONS

The single biggest thing you can do to build differentiation is ask the right questions for your business. Unfortunately most start-ups begin by asking the *wrong* questions, seeking answers to the wrong problems. They end up selling some knockoff product that's a few dollars cheaper or has a couple of extra features. It doesn't cut the mustard with either customers or investors.

At my Montana ranch there are several months each year where it gets very dry. The entire ranch and surrounding area of several million acres are fire-prone. The local fire station is great, but driving ten miles to the ranch could take twenty to thirty minutes. By that time we would have a big mess on our hands.

We wanted a fire truck. We started off by asking the wrong question: "Where can we buy a used fire truck?" We quickly learned that we could spend lots of time and even more money finding a used truck.

So we asked a different question. We backed up our focus a little bit and asked ourselves, "How can we get a fire truck on the ranch?" With that question we realized we didn't necessarily need to buy a used truck. How about converting one of our existing ranch trucks into a fire truck?

That turned out to be the right solution. We bought a hundred-foot hose, an old two hundred-gallon tank, and a Honda pump. We rigged some ingenious plumbing and—voilà—a fire truck! To maintain the truck as a working part of our existing fleet, we made the changes temporary. We mounted the equipment on a base that can be lifted off and on the back of the truck.

So far we've used it to put out three fires. Each time we were done putting them out by the time the local fire trucks arrived (and they are very prompt). All they had to do was clean up.

Asking the right questions during the product planning also drives differentiation. Greg Gianforte of RightNow Technologies started off by sending the first draft of his one-page product specification to twenty potential customers. He made follow-up calls and set appointments to meet with them to talk about his idea. At the meeting Greg asked open-ended questions, allowing his customers to tell him what was important to them.

1. What is going to earn you the biggest bonus this year?
2. What are you losing sleep over?
3. What do you need to accomplish your job?
4. What are your challenges?

Notice that he did not ask obvious questions like "Do you want to buy my product or service?" or "What is the product or service you need?" It's Greg's job to figure out those answers himself, creating a product or service based on the deeper, more useful information he gets with the answers to the other questions.

THE 10X OR BETTER RULE

A young Technoid team came out to Entrepreneur America once. They were making a very high speed router that could be used by Internet operators, telephone companies, or end-user customers to send data at very fast speeds. Right off the bat they violated the 10x rule.

The founders started their product planning by asking, "How can we build a router that will sell for $30,000-$50,000 with all the same functions as Cisco's $100,000 router?"

But that target is aimed too low; the company isn't being aggressive enough. Cisco's base router was $50,000, and a few extra bells and whistles brought the price up to $100,000. So Cisco would have blown them out of the water in sales pitches, "Our router is $50,000 also, plus it's Cisco . . . blah, blah." The other problem is that building products always costs more than the original projections. A

$50,000 router quickly becomes a $75,000 router, and then you're selling on a tiny difference. On a difference that small, 15 to 20 percent, the established competitor will always win.

The key is to offer better functionality *and* better price— ten times better. This company finally settled on the right question: "How can we build a router that sells for less than $10,000 and includes strong features?" If you're ten times better at a lower price, you'll blow the competition away.

THE ONE YEAR OR LONGER HEAD START

A young MIT start-up I worked with, Iridigm, created a flat panel display for mobile information tools. Called Digital Paper, it has the look of high-quality printed media like a book or magazine but is dynamic enough to display full-motion video. It enables products like featherweight wireless Web pads that can be read in any lighting condition and run continually for more than a day.

Iridigm has established itself as an expert in harnessing the new field of micro-electro-mechanical systems (MEMS) for information display. It has done so by combining this new area with the mature field of LCD thin film process technology. Our extraordinary performance advantage derives from replicating the beautiful iridescent colors seen naturally in butterfly wings and peacock feathers. Iridigm's micromachine structures mimic on glass the effect seen in butterfly wings.

Digital Paper exploits existing LCD production lines and components. All attributes of this new technology have been demonstrated, and it has been broadly patented. The development of the technology and assembly of the core team was supported by the U.S. Department of Defense and took three or four people over four years. Now that's a serious head start.

Iridigm's goal is to be the mobile face of the Internet for

users all over the world. With their broad head start, they have a good shot at achieving the goal.

THE IMPROVING WITH AGE RULE

None of my start-ups have aged enough to prove this rule yet (although I'm sure several will). At Ascend, one of our differentiators was our "any to any" ability. That means that our networking boxes supported any popular phone line (ISDN, dial-up or T1) and would communicate with other boxes anywhere.

This second feature, communicating with other boxes across the country, only improved with time. Ascend was always adding new phone lines as they became available. So our box using dial-up phone lines in Podunk would talk to a box using T1 in Los Angeles. We added new line types as they became available, making them work with the old ones. As the product aged, we raised the bar for our competitors.

Spreading the Sticky Virus

VCs and savvy investors talk a lot about "scaling." They're hot for companies that have a natural fast-growth pattern and a business model that will support the growth without incurring additional costs. How does the business take off once it hits a customer? Viral scaling tends to spread quickly to other areas within the company. It spreads to the sales channel almost by itself, back to the supply chain and to anyone in contact with the customer.

To tell if your model scales, look for a natural growth pattern—your product or service will reach new potential customers naturally, as it's being used. Your company has the resources to handle the increased sales with minimal new hires and little additional training or increased development. Serious scalability is viral, meaning that it spreads quickly among customers, suppliers, wherever it makes contact.

Here's an example, the start-up I mentioned earlier that does on-line training for the automotive industry. Vellis's scalability is impressive.

First, the model ramps up without additional costs. Once Vellis creates the applications and content for one training course, all the technology needed is already in place for subsequent courses. The employees of Vellis's customers are familiar with how the training system operates. When the company comes up with training needs in other areas, they ask, "Can your technology handle that if we provide the content?" So the product spreads laterally within the company, while all Vellis has to do is add more content. The system scales effortlessly within the customer's company, from product line to product line.

It also scales through the sales channel and through the supply chain. For example, the people who *retail* the trucks will be brought into the Vellis fold with the same training, accessing the same on-line information. This begins to create a training course that can be used at almost any other company. The companies that supply parts for trucks, the service companies that warranty the engines, manufacturers that build the engines—the scaling just keeps pushing up the supplier chain. See what I mean about viral scaling?

Vellis discovered an added bonus—once they've built a Web portal for their customers (whose technicians use it to look up information for servicing trucks, for example), why not add a business-to-business sales component? An employee looking up information on fixing an engine can click on the parts or tools, ordering them easily. Vellis's technology handles the order processing, earning the company a small piece of every transaction. The company can partner with retail tool companies and add their inventory to the site, taking a piece of their sales as well.

Vellis has big potential for viral scaling. As the word implies, the model spreads out to new accounts like a virus. Once Vellis embeds its system in a big company such as Mercury, it begins to take off in two directions: toward the sales channel and dealers and toward the supplier chain.

Plus, the model is very "sticky," which is something VCs like to see. It means that the product or service gets deeply embedded within the customer's company, making it difficult for the customer to change to a competitor's product. Once Vellis gets its customer's employees trained on one process, whom do you think they're going to call when they have another need? They're not going to invest in learning a whole new training system. As long as Vellis can work with the client's data and provide new training modules, they're in. It's this kind of relationship that will keep your customers partnered to you for life.

Vellis is ready to talk to investors. But even companies like Vellis, ones that lock up the gold mine with the key elements I've talked about, risk flubbing things. The next step is learning to run the investor meeting, which can actually be more critical than how well your company is put together. In the next chapter I'll tell you about VCs that have walked away from no-brainer investment opportunities because the entrepreneurial team blew the meeting. Make sure your company isn't one of those.

EXERCISE:

Value Questions

1. What corporations need your product and why? Be specific, focusing on one company as an example to answer the rest of the questions.

2. Who would be the decision maker for buying your product or service?

3. What is going to make him or her a "hero" by making that purchase? How does your product or service relate to that?

4. What budget would your product or service's purchase come from?

5. How much money does your product or service save or make this person? How much money will *you* make?

Differentiation Questions

1. List the red polka dots of your product.

2. List the questions you asked in developing and building your product.

3. List the major features/functions that are 10x better.

4. How have you established your year (or longer) head start?

5. How will your differentiation improve with age?

Scale Questions

1. What triggers a "feeding frenzy" for your product?

2. What will make it spread to your customer's supply chain?

3. How will it spread into the sales channel?

4. What makes it spread to other groups within the company?

5. What pushes it to spread into other applications?

Sticky Questions

1. What prevents your customer from throwing you out on fairly short notice?

2. How entwined is your product with your customer's "backroom" and "front-room" internal systems?

▶ You are welcome to refer to the Entrepreneur America Web site for additional information (www.entrepreneur-america.com).

5 *Peeing in the Wells*

ROAD MAP: ▶ Your presentation is polished, it's finally time to meet investors. The key is to carefully target the right VC groups. Otherwise you'll start hearing "no" answers, and one "no" tends to turn into many.

Last winter, in the middle of Montana's first snowstorm, an old friend of mine from Boston came out to see me at the ranch. It was actually a professional visit, and he brought a few other people with him—the rest of his management team. My friend, Dick, wanted my advice about his start-up, called Everfile. The company had developed Web-based document-sharing technology, and it looked good. But as they talked to me about the business, I got frustrated.

"Listen, you've already screwed up," I told him. "It's so bad that I don't even know if I can help you." What Dick's Technoid Wanna-be had done was to spray the business plan to dozens of VCs across the country. By the time he came to see me, Dick had run to every investor in a friend's Rolodex and had been turned down by all of them. His business idea wasn't all that bad, but his presentation was boring. I almost dozed off as one of his engineers stood in front of the whiteboard and mumbled about servers and encryption protocols. I couldn't believe that's what they had gone to VCs with.

But he is an old friend, and I agreed to work with him.

After a weekend of blunt words from me, the team drove away in the snow with strict orders to reposition their business plan and hone the presentation.

Like my old friend, many Wanna-bes call me for help when it's almost too late. It's very hard to regroup once all the venture capitalists within earshot have heard their cockamamie stories. The start-ups that call me before they target VCs have much better luck getting funded.

It all leads to the single biggest mistake made by entrepreneurial Wanna-bes—"peeing in the wells." Money is the water that keeps your company alive; without it the company will wither and die. But if you start your search the wrong way, you risk ruining your entire water supply. It could take years before the water is clean enough to go back and try again. You end up in a problematic position—knowing *too* little, moving to get funding from professionals *too* early. And the result is *too* bad.

Clearly the best idea and the sharpest team won't get you anywhere if you don't have the money to build your business. But that's not the only reason to polish your pitch. The lessons of this chapter apply to more than just fund-raising. A good presentation tells the story of your company in such a compelling and effective way that people want to hook up with you. Using these techniques, you'll land more customers, recruit better executives and employees, grab the media's attention, and raise money (from VCs, angels, banks, or other sources).

When I talk about seeking funding, I should point out that some people think I'm hopelessly out of date with my four-step dance approach. They argue that it's a "new funding landscape" out there, and things have changed with VCs in the past few years. They move more quickly now, without requiring as much due diligence.

Well, it may be a new landscape, but I'd rather camp my start-ups under a mature shade tree than some scrawny little shrub. I admit that there are a large number of new VCs with fat funds ($300 million to $1 billion). But they've only recently jumped in the game. They're untested in building a company and surviving bad times. They've only been working in happy and rich dot.com times; who knows if these little shrubs will make it through a downturn? The other risk here is that they're rushing to fund some pretty miserable business models. Sure, they might be willing to take a chance on you, but they're not likely to discover the fatal flaws in your plan or team.

That's not the kind of company I want to help build. Because I'm interested in growing solid, sustainable businesses, I work only with established funds—the "elms" of the industry. My VCs might be tougher to land, but it's because they're taking the time to determine whether or not you're worth it. When Wanna-bes manage to secure one of the investors I target, we are fairly certain that it's going to be a long-term relationship. And to date, half of the dozen Wanna-bes of which I'm on the board either have had or are heading toward initial public offerings. I think that's proof enough that my approach works.

Starting Your Engines

There are only two things you need to go out and ask for money. One is a single-page executive summary; the other is a team presentation that summarizes all the hard work and research that went into the business model. These are the two things I work on out at the ranch. We practice the presentation relentlessly, and we edit the executive summary to weed out all the meaningless marketing phrases. By the time we're finished, my start-ups have a one-pager they can mail to potential customers and VCs who take their calls. And they have a polished presentation they can give if the one-pager wins them a meeting.

Okay, there are really three things: you also need a list of references. And for extra credit, develop a five- to ten-page written version of the presentation. If potential investors like what they see, they'll want to check up on your references and pass along the summary to partners who couldn't make the meeting.

EXECUTIVE SUMMARIES THAT SELL

The ones from most of my Wanna-bes stink (at first). The idea is to create a one-pager that will get you in the door for a VC or customer meeting. Let me show you a good one and explain what makes it work. This one is from a prospectus of a company called FreeMarkets, which I haven't worked with. It's one of the best I've ever seen:

> FreeMarkets creates customized business-to-business on-line auctions for buyers of industrial parts, raw materials, and commodities.

Big score with the opening sentence! It's a clean, brief, and (most important) easily understandable statement of what the business does. They've skipped the gobbledygook of buzzwords and hyperbole that most companies indulge in.

> We created on-line auctions covering $1.0 billion worth of purchase orders in 1998 and $1.4 billion worth of purchase orders in the nine months ended September 30, 1999. We estimate that the resulting savings for our clients ranged from 2 percent to more than 25 percent.

Note that these hard numbers are customer driven. That generates instant excitement about the potential for this business. What if you don't have sales yet? Talk about prospective customers instead; for example, "ABC company plans to use our product for Y, which would result in X million in the first year and Z million in the second." This means you will have had to talk to customers instead of quoting some vague report.

The last sentence of that paragraph—"We estimate that

the resulting savings for our clients ranged from 2 percent to more than 25 percent,"—is a great statement of value proposition. It's terse, believable, and powerful. Clearly this team has a shot at building a real company, because saving customers money always means you'll make money.

> Since 1995 we have created on-line auctions for more than thirty clients in over fifty product categories, including . . . [industry specifics] . . . More than two thousand suppliers from over thirty countries have participated in our auctions. Our current clients include . . . [pretty impressive Fortune 500 names]. Two of our clients accounted for 58 percent of our revenues during the first nine months of 1999.

This paragraph paints an excellently detailed picture of the market and customers. Don't have customers yet? If you've followed my advice, then you'll at least have beta testers at this stage, perhaps even small revenue figures to talk about.

> Based on industry research and government statistics, we estimate that manufacturers worldwide purchase approximately $5 trillion each year of "direct materials"—the industrial parts and raw materials that they incorporate into finished products. Because these direct materials are often custom-made to buyers' specifications, there are no catalogs or price lists to enable buyers to make price comparisons. The process of purchasing direct materials is further complicated by the fragmentation of supply markets and the importance of product quality in supplier selection. Because this complexity leads to market inefficiencies, we think that buyers of direct materials often pay prices that are too high.

Here is a strong statement of why this business exists. It summarizes the unmet need and the industry forces that combine to create the business opportunity. All too often Wanna-bes shorten this statement to refer simply to some "zillion dollar" number, explaining neither what specific opportunity exists in that market nor what industry forces create the opportunity.

> The Internet offers an opportunity to create more efficient markets for direct materials. As the number of Internet users has grown, large companies have increasingly adopted electronic commerce as a way to do business . . . [some market research stats on Internet market growth] . . . we believe that Internet technology alone cannot solve the problems faced by large industrial buyers. To solve those problems, we think that Internet technology must be combined with services that are customized to buyers' needs.
>
> We combine our proprietary BidWare Internet technology with our in-depth knowledge of supply markets to help industrial buyers obtain lower prices and make better purchasing decisions. In a FreeMarkets on-line auction, suppliers from around the world can submit bids in a real-time, interactive competition. Our auctions are "downward price" auctions in which suppliers continue to lower their prices until the auction is closed. Before each auction, we work with our client to identify and screen suppliers and assemble a request for quotation that provides detailed, clear, and consistent information for suppliers to use as a basis for their competitive bids. Our service, which we call "market making," creates a custom market for the direct materials or other goods or services that our client purchases in a particular auction.

This section starts off with the normal "Wow, what a big opportunity we have here!" It's the last sentence of that first paragraph that's the real grabber: "To solve those problems, we think that Internet technology must be combined with services that are customized to buyers' needs." It hints that more than just Internet technology is needed for the business to succeed. That sets up the next paragraph, which is a detailed but clear statement of how the company's business and technology actually work. Plus it reinforces the value proposition stated earlier.

> We seek to be the world's leading provider of business-to-business on-line auctions. Our strategy is to extend our

client base in our target market of large purchasing organizations. We also intend to expand into additional product categories where our on-line auctions can generate savings for buyers, and to add new functions and features to our BidWare technology to further automate portions of our market-making process.

The perfect "rah-rah" ending gets the investor whooped up about the opportunity. It also propels the reader forward, toward a future vision. This is a company that's going to take you somewhere!

NAILING THE PRESENTATION

Okay, you've scored your first VC meeting. A partner at the hottest VC firm in your industry is meeting you for breakfast. So what do you do? You flub it.

Here's what I see all too often: Wanna-bes who don't prepare. No matter what category they're in, they usually haven't built a prototype and tested it. In fact, they haven't talked to customers. They dash in and launch into their hundred-frame Powerpoint presentation. They drone on and on, not asking questions or interacting with the potential investors (or me, if they're out at the ranch). When someone does pipe up with a question, the Wanna-be gets into an ideological argument about the impressive attributes of the product—not the market or potential customers. When the whole event is finally over, the investors shake hands, smile, walk out, and never call.

One of the best presentations I sat through was from a company called Creditland.com. The guys summed up their entire operation on one slide. That was it! It was a meaty little slide that showed the basic structure of how the business would work. The team talked a little bit about applications. It was terrific because they didn't sling around a bunch of buzzwords. I admitted that for the VCs there would have to be a few more slides on competition, barriers to entry and financials. But it was a great beginning because it was terse.

The presentation is more than just some pretty slides. It's an intensely boiled-down version of your business model, with answers to all the tough questions like Who are the customers? What is the application? What is the value proposition? How do you make money, and how much? How do you differentiate yourself?

Just as any great story is art, there's an art to a great business presentation. Basically you're telling the story of your company. To do that, you must know your material, be articulate, and be able to convey the message simply and thoroughly. This is not an option. You don't have to be a great public speaker, but you do have to be convincing in a one-on-one situation. Your presentation must always segue cleanly from one slide to the next. If you're no good at speaking, accept lots of speaking engagements and overcome that fear. I got professional coaching once I realized that there was a direct correlation between the success of Ascend and my own ability to communicate.

My Web site (www.entrepreneur-america.com) has a standard outline for a brief slide presentation that I teach my Wanna-bes, using Ascend as an example. When I say "slide," I mean flipbook page, Powerpoint screen, whatever you're comfortable working with. I use a flipbook. The slides are really just a prompt for your speech; alone they have little or no information. None of them should be wordy or busy.

The first four slides in your story make or break the presentation. This is where you announce what the problem is and how your company solves it. They should take only three or four minutes, a very crisp pace to hook the audience.

Slide 1: What you do: Set up the presentation and grab the audience's attention. Take twenty seconds to introduce yourself and the team.

Slide 1

What the audience looks for: A *team*. Is it representative of the skills needed to pull off the start-up? Does it operate like a team or just a single presenter with the followers? I am not interested in titles. Teamwork is the number one item that VCs and I want to see executed well.

Slide 2: What you do: This "opportunity" slide is your chance to explain what need your business will fill—*not*

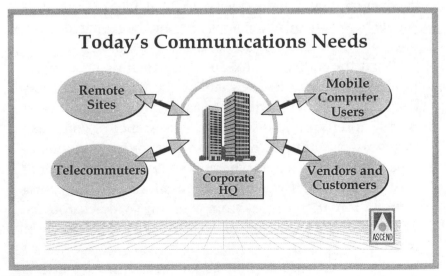

Slide 2

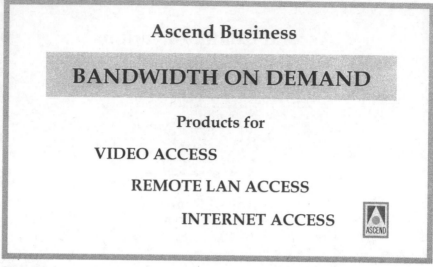

Slide 3

your product or service. Take about ninety seconds and move quickly to third slide.

What the audience looks for: A clearly defined, compelling need. Your customer should be saving lots of money and improving performance.

Slide 3: What you do: This "business" slide slams home the point of what your business does. It's okay to talk about product families, but don't get into the details of individual products yet. Spend about a minute, and be prepared to answer questions about customer use at this point.

What the audience looks for: A compelling need—your product or service is essential for the customer to function. The customer is willing to pay dearly because it saves money and time and is strategic. The essence of your business must be captured here in a short, pithy, and memorable statement. Here is where to show a prototype, or at least be able to answer detailed questions about it, using information that comes from working with customers. The greater the customer interest, the greater the investor interest.

Slide 4: What you do: Talk about "market size," which is

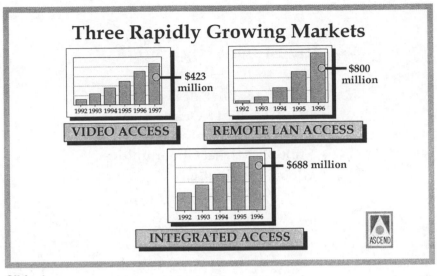

Slide 4

best illustrated from the bottom up, by talking about customers again. Spend half a minute saying something like "Let me tell you what Citicorp is doing with our product. This single account spends $X [average sales], then it buys $Y each quarter or each time some event occurs." The audience will fill in the rest.

What the audience looks for: An example that shows how much the customer typically buys and at what price. What are the continuing sales? How is this customer typical or representative? It must be clear that you've talked to real customers.

Slide 5: What you do: Step back and spend half a minute talking about the highlights of the company, including employee size, revenue size, products status, and markets served.

What the audience looks for: A cohesive story with the company positioned where it should be.

Slide 6: What you do: A thirty-second segue into what business areas you'll be covering in the presentation, pointing out that you'll get to everyone's questions eventually.

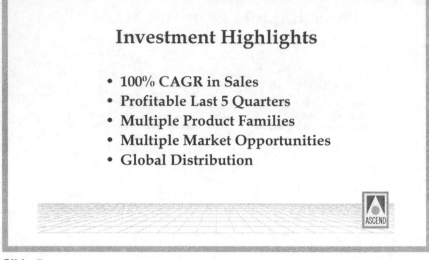

Slide 5

What the audience looks for: A clean transition and good pacing. Practice, practice, practice!

Slides 7, 8, 9: What you do: Tie industry forces with your business. Then close the loop on how/why your products play into industry forces. Do, at most, three industry forces slides, taking four or five minutes for all three. If a previous slide already touched on this, skip it here.

Slide 6

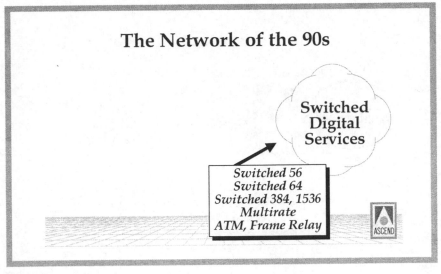

Slide 7

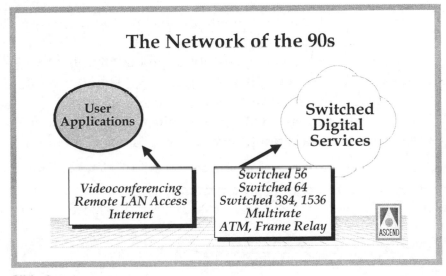

Slide 8

What the audience looks for: Compelling industry forces and a product position that exploits them. Also, the presence of serious and sustainable driving forces. I listen for the reasons why those driving forces will continue to be present. The best businesses are monopolies where the compelling force lasts essentially forever. The stronger the

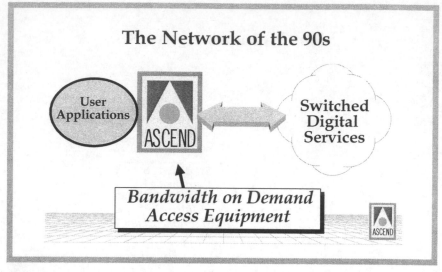

Slide 9

link between industry forces and product position, the easier it is to achieve customer traction.

Slide 10: What you do: Talk about product families and their market focus. Refer back to chapter 3, "The Sunflower Model," for help in putting this slide together.

What the audience looks for: Products. If you have an idea, then build and test it on customers. If you can't build

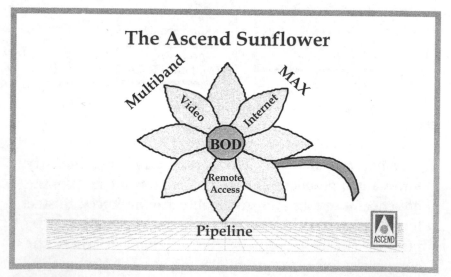

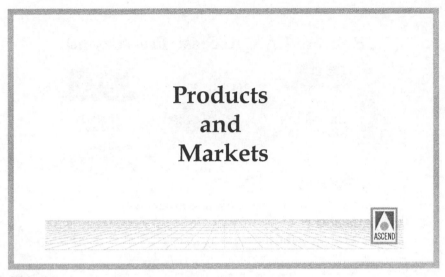

Slide 11

it (no money?), you can still workshop the idea with real customers.

Slide 11: What you do: Take about five seconds to shift gears. Announce that you're going to talk about how customers did things before you and how they now do it with you. Also, you'll get to the product status and a real live customer vignette. The next four slides do exactly that.

What the audience looks for: Clean, smooth transition. A polished team that works together.

Slide 12: What you do: Spend a minute explaining what customers used to do, before your company came along, and the drawbacks to it.

What the audience looks for: Serious drawbacks in the old model, ones that impact the company's bottom line and its strategic ability.

Slide 13: What you do: Spend another minute showing how your application specifically solves the previous problem, measurable in dollars.

What the audience looks for: A "hard dollar" payback in the new method. Paybacks are measurable either in hard

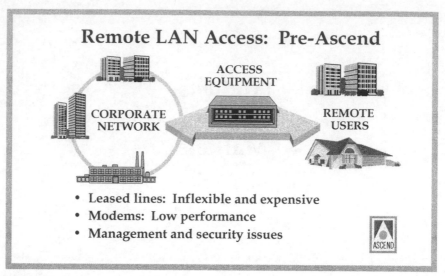

Slide 12

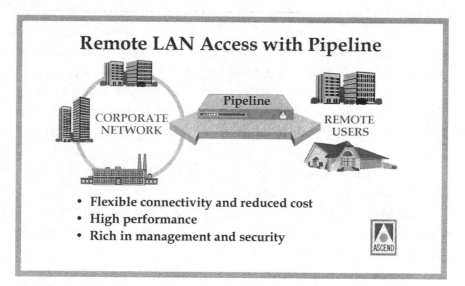

Slide 13

dollars, or they take a leap of faith ("Our product is easier to use than others"). If it's not truly quantifiable, few people will care.

Slide 14: What you do: Mention where the product is, in development. Spend half a minute on this.

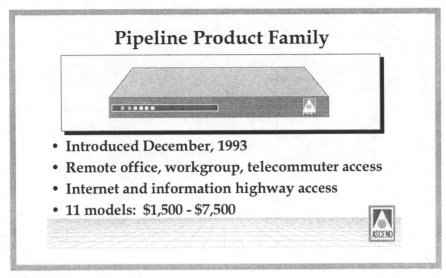

Slide 14

What the audience looks for: A prototype that's done or nearly done, with two or more applications for the product or service. Personally I keep an eye out for ones that have substantial core technology and are a bit "contrarian" (haven't been done before). I also look for major differentiation.

> **TIP**
>
> When explaining a complicated concept, start off by using an example. Then explain the actual idea, following up with another example. People grasp ideas more easily if they're related to something familiar, like "Ascend builds dial-up networking equipment. For example, our boxes are the plumbing of the Internet."

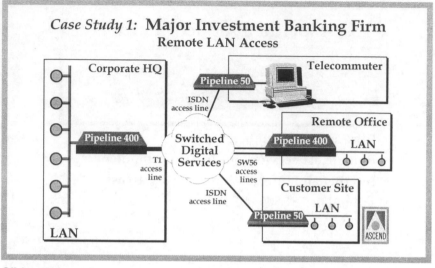

Slide 15

Slide 15: What you do: Talk at length about how customers (by name) will use your application, why they chose it, and what the savings are. Spend three to five minutes on this meaty topic.

What the audience looks for: Customers. You will *not* raise money without customer references.

Slide 16: What you do: Spend five seconds transitioning into the next topics—product differentiation and competition.

Slide 16

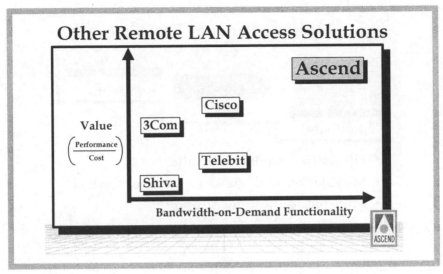

Slide 17

What the audience looks for: Teamwork, smooth transition.

Slide 17: What you do: Give a five-minute talk about what you do differently and how it gives you an edge. Explain why the competition cannot duplicate this easily. You might want to add a subsequent slide that goes into more detail on competitive differentiation.

What the audience looks for: A complete understanding of who the competitors are; a rational, objective, and measurable argument of why you will win; and breakthrough, contrarian products—not "me too" products.

Slides 18 and 19: What you do: This is the beginning of your competitive advantages. Slides 18 and 19 show two advantages I talked about regarding Ascend.

What the audience looks for: Real advantages that build competitive barriers of at least nine to twelve months over competitors.

Slide 20: What you do: This is where you talk about the future.

What the audience looks for: A real business that has

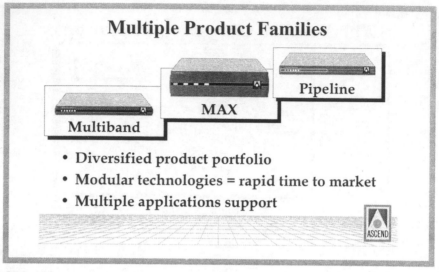

Slide 18

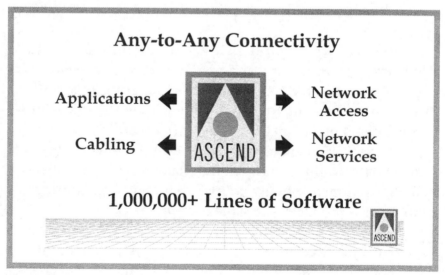

Slide 19

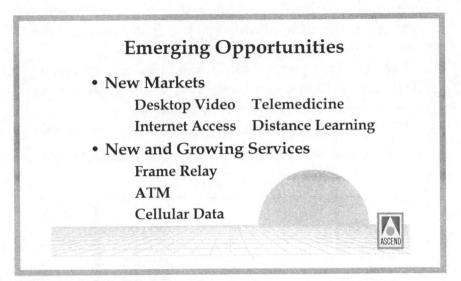

Emerging Opportunities

- **New Markets**
  Desktop Video   Telemedicine
  Internet Access   Distance Learning
- **New and Growing Services**
  Frame Relay
  ATM
  Cellular Data

Slide 20

many opportunities to leverage its technology into multiple markets and products.

Slide 21: What you do: Segue into the next group of slides—financials.

What the audience looks for: This is probably the most misunderstood section in every business presentation. The really interesting part of the financials is the reasoning *behind* the numbers. The audience wants to see, operationally, how you're going to get a piece of the market. In these slides it's not the graphs that will help you, but the story about how you got the numbers. I know no serious,

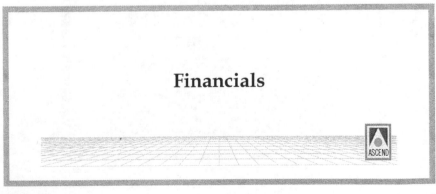

Financials

Slide 21

successful investor who puts much store in a spreadsheet—
these numbers must be driven by real customers. Take
about five minutes for the entire set.

Slide 22: What you do: This is your pro forma revenue;
remember you built this from the bottom up based on a set
of sales assumptions. Prepare a five-year outline.

What the audience looks for: Reasonable, conservative

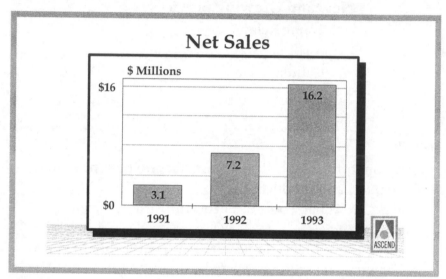

Slide 22

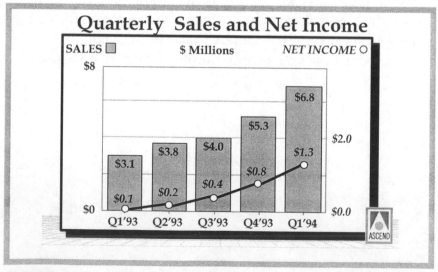

Slide 23

revenue numbers. Plans that show $1 million, $30 million, $80 million, and $200 million generally spend accordingly, and of course the spending stays on plan, but the revenue rarely does. So keep the numbers conservative.

Slide 23: What you do: This is your pro forma sales and net income by quarter. You should prepare a two-year (eight-quarter) projection. Talk about how the numbers

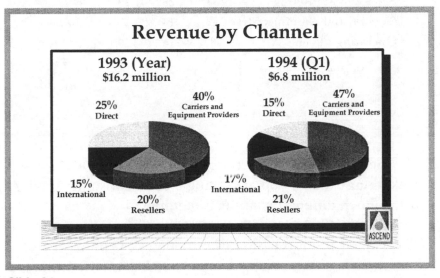

Slide 24

were derived. Make sure that you stick to the norms unless you have a powerful reason.

What the audience looks for: When the break-even is reached, measured against the norms for your industry. For example, network equipment companies take twelve to fourteen quarters, spending about $10 million to $12 million on a $16 million to $20 million run rate at break-even.

Slide 24: What you do: This is pro forma for sales by channel. Remember in building your bottom up sales plan, you will have made assumptions on whether you are doing direct sales, OEM, VAR, etc.

What the audience looks for: A rational sales distribution plan. For example, if you are selling a $100,000 com-

Financial Model

%% of Sales

	Q3'93	Q4'93	Q1'94	*Objective*
Gross Margin	65%	65%	65%	*64-66%*
Research and Development	12	10	9	*10-12*
Sales and Marketing	33	28	29	*26-28*
General and Administrative	12	13	7	*6-8*
Operating Income	8	14	20	*Low 20s*

ASCEND

Slide 25

plex network carrier box, then direct sales and OEM sales makes sense, but if you are selling a $500 consumer product, a much different channel is required.

Slide 25: What you do and what the audience looks for: This is your pro forma financial model. Your industry has norms. Learn them. A good source for information is annual reports. If you depart from the norms, have a good reason.

Calculate the percentage for gross margin, research and development, sales and marketing, general and administration, and operating income by quarter (actual or pro forma). Then estimate, based on industry norms, the same percentages going forward. This becomes your objective.

Slide 26: What you do and what the audience looks for: Recap—about thirty seconds of hitting your high points.

Slide 27: What you do and what the audience looks for: Ask for the order. Either make the sale or get constructive feedback at this point; don't leave the room without one or the other.

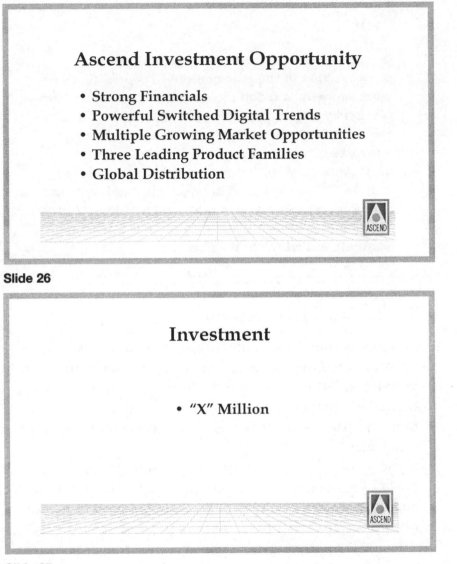

Ascend Investment Opportunity

- Strong Financials
- Powerful Switched Digital Trends
- Multiple Growing Market Opportunities
- Three Leading Product Families
- Global Distribution

Slide 26

Investment

- "X" Million

Slide 27

THE ONE-SLIDE PRESENTATION

As with my friends at Creditland.com, it pays off to have a one-slide presentation ready to go. Don't cut corners on your information—just pare it down to summarize your company's market, product, value proposition, differentiation, and barriers to entry. It's a lot to cover, but you need it to make the telephone pitches.

Here's how it usually works. You, or someone you know, pitches your company to a prominent VC. He's interested and asks for a business plan or slides. I don't give those out, usually because most of my entrepreneurs can't write well enough. Also, I don't want the plan wandering around out there without the entrepreneurs attached to it—I want my team to make the pitch, then get either a "yes" or a "no," and "why." Otherwise we're wasting our time sending the plan around.

Instead I ask to set up a call. That's when we use the "one-slide" technique. From that simple slide (by looking at it on our end and e-mailing/faxing it to the VC), we can make a quick but thorough presentation over the phone.

The discipline of creating the slide forces you to summarize your entire concept in as few words as possible.

The Waterford Experience: When Financing Is a Crystal-Clear Outcome

The approach I teach out at the ranch is careful, deliberate. We stalk investors in a well-planned military campaign. We target them methodically, one at a time. The point is to get them on your side so they want to help fix the holes in your plan, instead of using them as an excuse to turn you down.

Virtmed is a Guts and Brains Wanna-be I worked with that had what I call a "Waterford experience." They followed the Entrepreneur America plan perfectly, with funding clearly in sight at the end of the process. I met the founders at MIT, where I was a keynote speaker at an entrepreneurial conference. Virtmed's product is a menu-driven PalmPilot application to help doctors track patient billing records. The Pilot links to a docking unit at the nurses' station that synchs with hospital accounting records.

The company carefully followed my advice on product development and testing based on customer feedback. They developed a beta product, which they were testing at Massachusetts General Hospital, in Boston, when I invited the team out to the ranch.

While they were in beta test, they started getting calls from interested VC partners. Fortunately they were religious about following one of my rules: When you're not ready to talk, tell the VC people to wait. It's hard, but it shows discipline. And you don't risk polluting the wells with a half-baked business plan. The Virtmed founders called me up every time they got a phone call, saying, "We really want to talk to them, but we told them we weren't ready yet." Instead they pumped the VC people for information on what investments they were making and what they were looking for.

When they came out to the ranch, we worked on the business presentation and two-page executive summary. The team practiced the presentation, then practiced again. Virtmed was relentless. They got so polished that even I was impressed.

At that point in a start-up's development it's time for the mentor to call one or two carefully selected VCs. The key is to focus, not scatter-shoot. Stick with one targeted investor until either 1) they agree to do the deal; or 2) they give up on it. If they give up, find out why and fix that problem before setting up meetings with the next VC group.

With Virtmed, we lucked out. We made a single phone call that paid off, to Chuck Neuhall of New Enterprise Associates. Neuhall chatted with the team for a few minutes and then agreed to see them in person.

After the meeting, Neuhall told me how mature the Virtmed team was. They weren't arrogant or defensive. They listened. Neuhall really liked them, and that's what helps sell a deal. NEA knew that they could work with this company's founders and bring in a good, strong CEO and management.

Neuhall was interested enough to assign another partner to do some due diligence on Virtmed right away. When a VC assigns someone to this, they're taking the deal pretty seriously. I told Virtmed to work with NEA cooperatively and quickly, providing all the information they requested. We gave NEA an exclusive for two weeks, telling the other VCs who called that we'd talk to them later.

While NEA researched Virtmed, it began to expose holes in the business model, which Virtmed was able to help fill in. So not only was Virtmed's business story getting stronger, but Neuhall was getting more and more serious about investing in it. Several weeks went by, and finally NEA agreed to offer $150,000 of bridge financing (a little of that was my money). Not a whole lot of money, but the fact that a top-tier VC forked it over gave us the credibility to look for others to complete the round and add management.

We set out looking for another partner. Because the only VC group we targeted had invested, several others were very interested. But for this round we wanted money *and* management. We had enough leverage to tell the investors that the first group who found us a CEO or VP of marketing could buy in. Frazier & Company of Seattle did so and completed the round.

Virtmed emerged with not only the required money, but also key management. When the time comes for an IPO, you can bet that investment bankers will take this start-up very seriously.

If a company makes it through, say, three steps of this process and still doesn't have funding, it means one of two things: they aren't doing their homework on the VCs (targeting firms or partners who don't specialize in that particular field), or the company just isn't incorporating the feedback into the business plan, so the new VCs are seeing the same old problems.

Maybe those business problems aren't answerable. They just can't be solved. In that case I tell the Wanna-be that it was a nice try, but it's time to get on with real life. If the founders want to ignore my advice and stick with the plan, the only way they'll get anywhere is to start tapping angels or family. At that point they've hit enough VCs to establish that the idea is, basically, nonfundable.

Don't Lose on the "Soft No"

When meeting day arrives, fools rush in with a one-hundred-slide, two-hour presentation that bores everyone to tears. They pack each slide with too much information, then try desperately to explain it all. Little effort is made to gather feedback or incorporate it into the presentation before rushing off to the next meeting.

> ▼ **TIP**
>
> If you have a savvy friend (or mentor), ask him or
> her to sit through a mock meeting. When I worked
> with Virtmed, founder Steve Hau wasn't too sure of
> himself in front of investors. He videotaped the
> presentation and made the whole team watch it.
> Why not? The camera doesn't lie. It points out
> things you'll never notice while you're busy talking.
> Steve was amazed to see how many pauses his team
> left during the presentation. Even he yawned while
> watching the tape, so he punched things up a bit.

Face it, that well is peed in. This type of approach will
never win funding. What happens in meeting after meeting is that the investors smile, act happy, and walk out.
They don't come right out and say, "No thanks," but unless they assign someone to due diligence right away
(and when they stop returning your calls), it's a soft
"no." They're just not saying "yes," which is the same
thing.

Your business idea might be decent, but the slipshod approach to running the meeting reflects how badly you
would run a company. That's what chases away investors.

Get to the meeting area early with your prepared business presentation, take a deep breath, and calm down. Remember that you've got nothing to lose—either you're
going to generate interest that might lead to funding or
you're going to get valuable feedback from investment ex-
perts that will focus your company for later pitches.

GAUGING THE RESPONSE

My dictionary defines "meeting" as "to come together."
Note that it implies there's more than one person (or team)

in the room. The most valuable part of the meeting is when the VCs start to talk.

One thing I always watch out for is the "happy hand-shake." Are they glad-handing everyone as they rush out the door? Bad sign. What you want them to be doing is asking more questions. Get them going by asking, "How did you like what you heard?" Then you can work toward the commitment:

> Wanna-be: So, is this going to converge with what you're investing in?

> VC: Oh, we have a process. Once we agree we like it, other partners need to review the material. Then we have a meeting . . . (blah, blah).

> Wanna-be: Yes, I understand there's a process. But you guys here in this room, what do *you* feel about it?

Here's where you start to get the skinny. Listen carefully to the likes and dislikes (someone should take notes). You'll start to get a clue as to whether or not they're interested from which side the feedback falls on.

The responses at the close of the meeting can be confusing—when they smile and say they'll call, is that a good thing? No way, you're dead. Here's what they're going to say and how to respond to it:

We'd like to give you a term sheet.

Sounds great! But be careful. There are unscrupulous firms that freeze you by offering a term sheet (the one- or two-page summary of the proposed deal). Although you might offer an exclusive for a week or so (particularly if you're working with one of my top-tier firms), don't stop talking to other VCs just yet. The first one to cross the finish line with the check wins.

It's really too small (or big or whatever) of a deal for us.

Code word for "sayonara." Ask them what is *really* bothering them about the deal.

There is another hot deal ahead of you; we'll stay in touch.
Also a nice "no." Again, ask what's the real issue.

We haven't invested in this market yet.
Your mistake! You never should have been talking with this group in the first place. Find another that specializes in your field.

We're not wild about you and your team.
This one is really tough, since you can't always change people's personalities. VCs won't invest in people they're not comfortable with. Work on presentation skills and hunt around for a better fit. But also be straight with your-self—if you're not CEO material, then be open to the idea that your company might need a management team.

There's too much competition, too little return.
Beef up your presentation on the competition. How will you sail through it, what's your advantage? What's your differentiation and barrier to entry?
It's unclear how the business makes money.
Make sure your story is crystal clear on this point. If you're not making money today, spell out when you will and whether it's going to trickle in or be a downpour. Use the value proposition to demonstrate this.

How does the business scale?
VC likes to see a nonlinear model. That means they want to see an exponential ramp-up, the Nike curve. Be very clear about how many months or years it will take for this to happen and what the exact "trigger" event will be.

Sorry, we're not interested.
The flat rejection. Doesn't happen often, but there's not really a way to turn this one around. Try to dig out more specifics on why you were rejected—be insistent. You need

this information to improve your pitch. Then regroup and move on to the next presentation.

We want to set up another meeting; how about tomorrow?
Very good sign. Probably they want to bring in different partners or the associate who will start due diligence. They're trying to build consensus within their organization. Ask what they'd like to see at the next meeting, then work that into the presentation.

We'd like to assign someone to start doing some work.
The best response you can get—they're serious enough to research your idea and the market. Get their contact name and e-mail address; if you don't hear from them in a few days, get in touch. Promise an exclusive for two weeks (they have to work quickly, and you're free to focus on them instead of making other pitches), and promptly give them any information they ask for. Delays will only cool their ardor and delay your funding.

HANDLING FOLLOW-UP

All of these responses, except the flat rejection, will create some kind of work for you. If you've entered what I call "the Promised Land," when the VCs are doing due diligence, then your work has just begun. The research doesn't stop until the check is in the bank.

After the term sheet, there are a couple of weeks of "getting to know you" dancing. Your goal is to secure one person within the firm to act as your company's sponsor, someone who is willing to push the other partners to make the investment. Stay in touch with the partner you originally talked to, offering references and additional information. Once you sell the sponsor, he or she takes on the responsibility of building the case for the other partners.

While you're riding the due diligence bandwagon, follow up promptly on all open-action items (requests for information, contacts, and the like). Make sure someone keeps a list of those at all meetings so that nothing slips through the cracks.

Work closely with your sponsor—the partner or assistant assigned to your project. This person can't completely green-light your investment, but he or she sure can stop the whole thing. If you fail to win this person over, you're sunk. Meet with all the consultants your VC partner asks you to, but research them beforehand. If there's an obvious conflict (the consultant starts to look like a competitor), then (nicely) offer to meet with someone else. Be straight about your comfort level.

Set up meetings with the VC group. Moving into three or four subsequent meetings is a good sign. At each meeting get precise feedback on what the VCs liked and didn't like. Rewrite your presentation to emphasize the "like," and fix the dislikes before the next meeting. If you haven't fixed the bad stuff, defer the meeting.

Don't harass the investor. Don't try to bluff your way through answers. The best response is, "I don't know, but I'll get back to you." Then make sure that you do.

What I've learned is that our first few VC meetings don't usually lead to the Promised Land. More often than not, my Wanna-bes get a nice "no," at the first meeting. They dig for specific feedback, then use it to strengthen their company story for subsequent meetings.

When things go well, the VC will eventually send the term sheet. There's usually some further haggling over the valuation and timing of the investment. But unless you have a big problem with the partners or terms, you're home free.

How Much Are You Worth?

In this section I'm just talking about the first round of investment. If you close one round, you're probably going to get more, but by then you'll be familiar with the process. I'm also talking here about "premoney" valuations, which simply means the company's value before the prospective investment is made.

To really know your start-up's value, you've got to build some financials that are based on real accounts, or prospects. Then you'll come up with a plan that speaks to the realities of doing business.

For example, a sales plan (if you don't yet have sales) might lay out the expectation that by Q2 the company will have one large customer, with one hundred users at $1,000 per user per year. By Q3 it expects to have two more customers, for a total of three hundred users. Project this out for eight quarters. These costs should be built on feedback from talking with at least one (named) account.

Once you have the detailed sales plan, think about the direct labor costs associated with them—sales staff and their costs. Then factor in the extra indirect labor that will have to come on board—prospect management, engineers, and support staff.

Now you have a cash picture. What is that for one year? Take that number and add 25-33 percent for margin of projection error. Product deadlines get missed, employees quit, leases increase, and all of that costs extra money. The VC insists on that little cushion because nobody is perfect in their modeling. They expect your financials to be as dead

> **▼ TIP**
>
> Say it'll take six million to float your first year—should you ask for ten? My advice is always to wait and see if the frenzy will take you there. VCs rarely shell out more than a year's worth of financing, so don't ask for it. But if your opportunity is hot enough, they'll offer you more.

on as possible, but they also expect to see a little fudge factor at the end.

That's the one-year number with the slush factored in. You're not going to get much more or less than that. It takes about three months to raise money, so if you ask for six months, you'd have to turn around and start over by the time you got it. Anything more than a year is just too long for VCs to fund without some sort of reality check.

AVOID THE SEESAW

What sometimes happens in a second or third round is that companies run up against investors who are less sensitive to valuation. Your company might be worth $50 million, but if you talk it up enough, you can find someone willing to pony up $100 million.

Looks pretty swell, but don't do it!

What will happen is that you'll be expected to seriously ramp up performance to meet the unrealistic infusion. If you don't, then in the *next* round your valuation will fall. Your company's track record begins to look like a seesaw instead of a nice, clean ramp-up.

For example, here's the linear progression that Ascend's valuation followed:

First round: $6 million
Second round: $18 million
Third round: $40 million
Fourth round: $100 million
IPO: $150 million

In that second stage, we could have puffed ourselves up to command a higher valuation. But what if we didn't blow investors away with our performance? I've seen plenty of companies with valuations that seesawed like this:

First round: $10 million
Second round: $25 million
Third round: $90 million
Fourth round: $40 million
Fifth round: $20 million

By this point morale sucks and everyone's quitting. A company that organizes and manages for long-term performance and valuation doesn't have to ride so many ups and downs.

MORE THAN JUST MONEY

It's tough advice for Wanna-bes to swallow, but trust me: who the investors *are* means more than how much they give you. Picking partners is as important as choosing a spouse; don't rush in. The thing to remember is that there's smart money, and there's dumb money.

Smart money can introduce you to buyers, customers, and other investors. It can fill the company's empty management slots with good people. Smart money is well networked, well respected, and well versed and will be an asset to your company. Don't forget that your investors sit on the board. You want someone who can help take your company to the next level, not just feed it cash. Look for VCs who have a proven process that helps them select the best entrepreneurs year after year. Smart money is attracted to smart entrepreneurs, so do your homework before talking to smart money.

Dumb money is just money. Quickie and Send Money Wanna-bes are the most susceptible to it. One characteristic of dumb money is that it rarely wants to make the lead investment. That makes it easier to keep it off the board (leaving that space open for someone who is more helpful). If dumb money is all you can get, stall and see if some smart money gets interested. Eventually you may have to be pragmatic and take it. But try to leverage dumb money to make future deals with smart money.

I'll take less smart money just to get the right investor any day. If we have a choice between $12 million valuation from Kleiner Perkins Caufield & Byers and someone else's $15 million, that's a no-brainer. I'd go with Kleiner Perkins. It's all paper money at that stage. The point is to focus on what assets are going to convert the paper to real money. The answer is great board members with lots of connections who have done this over and over again successfully.

Another advantage of smart money is that once you've gotten financed, it's time to focus on managing the company (instead of looking for money). Experienced investors sitting on your board will help you battle competitors—something I talk about in the next chapter.

One thing I advise is to try to get at least two quality VCs in the first round. If you're raising $20 million, maybe it will take three VCs. The more investors, the more you're amortizing your risk and increase the likelihood of having available deep pockets for subsequent rounds. The subtler issue relates to your board. If there's only one investor on the board controlling your stock, personal dynamics can create real problems. Spread out your risks and rewards. If your company is going to grow successfully, you're going to be in bed with these people for many years to come.

EXERCISE:
Writing the Executive Summary

Once you've done the hard work of researching your idea with customers, then you're ready to get started.

1. Write the opening sentence. This is the crystal-clear statement of what your company does. Work on this single sentence until there's no jargon, it's concise and each word contributes to the idea (no filler). Here's the example I used earlier in this chapter: "FreeMarkets creates customized business-to-business on-line auctions for buyers of industrial parts, raw materials, and commodities."

2. Next comes the "how-sweet-it-is teaser." This sentence explains what your company has been doing to take advantage of its market. Example: "We created on-line auctions covering $1.0 billion worth of purchase orders in 1998 and $1.4 billion worth of purchase orders in 1999."

3. Now it's time to sum up the value proposition. This key sentence basically shows how your company will make money by way of explaining what savings the product or service offers customers. Example: "We estimate that the resulting savings for our clients ranged from 2 percent to more than 25 percent."

4. Next comes the "customer wow" section. Here you show either the sales results so far or (if you don't have customers yet) the hard work you've done with prospects. Go back and look at the FreeMarkets example earlier in the chapter for ideas.

5. This is the spot for the statement of why your business exists. This paragraph should position your company firmly within the market, explaining clearly the business opportunity and your model for taking advantage of it. Return to the FreeMarkets example if you need help.

6. The technical section follows. Here is where to explain your company's operations. Explain clearly what technologies or other strengths you'll be leveraging and how. Play up any proprietary property your company has.

7. Time for the finale. These final few sentences project the

company into the future, summarizing growth plans and showing investors that your company has an exciting, strong, and successful life ahead.

8. Now take a stab at the presentation. First, take a blank piece of paper and divide it into sixteen small squares. With a pen or pencil roughly outline these 16 slides. The advantage of this technique is that you can't write too much—the whole idea of the presentation is to keep the slides minimal and rely on your mouth. Once these are blocked out, you can take them to the computer and pretty them up. Do not add more information at that point!

9. Rehearse it, videotape it, refine it.

▶ You are welcome to refer to the Entrepreneur America Web site for additional information (www.entrepreneur-america.com).

6
Sucking the Air out of the Room

ROAD MAP: ▶ If your company is successful enough to grab investor attention, it's already on your competitors' radar. Don't let down your guard. Choose your plan of attack and lock up your market.

Several years ago two competing start-ups independently came up with the same product idea. One of them was Ascend, the other was Company X (run by a guy I'll call Bill), about six months ahead of me in product development. At a trade show we ran into each other and started chatting.

We stood next to Bill's hotshot display area, eyeing a clump of khaki-and-blue-shirt guys huddled by a computer monitor.

"See those people over by your chief engineer?" I said to Bill. "They work for me, and they're probably pumping him for information."

Bill smiled at me. "Oh, that's okay. It's complex technology, it won't make much difference if they just chat about it for a few minutes."

I smiled back. "Yeah, well, we're completing the same product."

He looked surprised. "Oh? Exactly the same?" Clearly he hadn't done his competitive homework on us.

"Not exactly," I admitted. "We're working on that and several other products."

He relaxed, saying, "Oh, then it doesn't matter."

"You're sure you don't want me to call off the piranhas?"

He waved his hand and smiled again. "Nah, this market is going to be huge—billions. We can share it."

I grinned. "Okay, but fair warning—I'm not good at sharing. I compete to win."

And win we did. Clearly Company X wasn't taking us seriously. They were sitting on their laurels, confident and self-assured. Well, six months later Ascend had raced ahead. We won their major customer right out from under them, and that credibility got us two other big clients in the industry.

Like many companies with a market lead, Company X had lost any sense of urgency they may have had. They had forgotten how fierce competitors could be. Instead of suffocating us as they should have, Company X just sat on their lead without working on any new products or focusing on any new customers. They basically waited for us to catch up. Now they're out of business.

A market lead isn't enough to fend off the hordes of competitors who will come at you. And they will come. You can't prevent copycats, they're part of the landscape of capitalism. You have a choice—you can cruise along as you always have, or you can go into "suck mode."

I'm talking about sucking the air right out of the room. Don't leave your competitors anything to breathe. In suck mode you make it difficult for established companies to recover and very expensive for start-ups to compete. This is what people don't like about Bill Gates. He sucks the air out of the room. Nobody can breathe when they're standing next to Microsoft, and eventually they just quit trying.

Recognizing Your Pivotal Point

Once you've secured funding, sales are beginning to soar and the management team is rocking, your battle has just begun. Ironically, the time to start worrying is when your company begins to look safe. That's when you're starting to show up on your competitors' radar.

This is a pivotal point in your company's growth, the moment that will determine your future. How you respond will define your company: will it be an industry leader, an also-ran, or a has-been? My advice is, if one of these things is happening, watch your back. Your pivotal point is fast approaching:

▶ Sales are beginning to really pick up. In fact, it's getting easier to get in and see customers.
▶ Customers know your name. At the very least, they've heard of you.
▶ Customers know what your product or service does, the specific area it serves.
▶ They think or know they need it.
▶ Larger companies are making noises about entering your market soon.
▶ Start-ups are sprouting.
▶ Medium-size companies are merging, planning to serve your market.
▶ Analysts are beginning to study your market.

The trouble is that if these signs are manifesting, it might already be too late for you to completely lock up your market. The best position to be in is ahead of the curve.

Suck Mode "On"

Every company would love to discover its own gold mine— a killer product, die-hard customers, or legendary service that keeps the money rolling in. It's possible to do but requires hard work. In mining you don't just wander around and wait to fall into some vein-lined shaft. You have to actually gather up your tools and do some prospecting. One thing I've learned is that the harder I work, the luckier I get.

There are several ways to blow away your competition: act with speed and surprise; build a better product; be a sales channel hog; smother your customers with service; be a moving target; and anticipate your competition.

THE NEED FOR SPEED

As a start-up you don't have the bottomless well of money and armies of employees that your competitors do. Good for you! You have the advantage because being small means you can move more quickly and use surprise. You can lurk in the shadows and strike.

Speed is the key. Moving quickly gets you to market first, buying time to develop it and get first shot at feedback from customers. They'll tell you what is right and wrong about the product, which you can then adjust. It's the second, revamped product that can give you control of the market. Competitors have no choice but to play copycat.

Speed is all-important, yet companies that try to move fast still manage to mess up. Instead of focusing on being first in the marketplace, they focus on *time to market*. New product ideas are sent tumbling down a tortuous path of multiple approvals and planning, marketing, engineering, and other management documents. Wimpy time goals are waylaid further by a focus on getting the product or service perfect instead of getting it out the door and then fine-tuning it.

Instead, ask yourself: What kinds of shortcuts can we take? What kinds of reductions can we make? Speed is the ability to return to the account and say, "You know what you asked me about last week? I have it." If it takes a year to get back to them, they won't remember you (and in the meantime they may have bought it from someone else who moved faster).

Set clear time goals, and educate the management team

about why they need to be met. I found the best way to do it was to send out my management team to talk to customers. A customer demanding a new product or service imparts a real sense of urgency that even the best CEOs can't duplicate.

Surprise goes hand in hand with speed. In the early days of Ascend, we used speed and surprise to snatch victory out of the jaws of defeat. It was 1991, and we were riding high with our Multiband product for video networking. Our customers included Compression Labs, VTEL, Sprint, MCI, and British Telecom, and we were pursuing Picturetel and AT&T. We had a letter of intent from AT&T. Then came the unexpected blow. We lost Picturetel to our competitor, Promptus.

At first we didn't believe it. We had the momentum; why shouldn't Picturetel follow the others? The answer was simple—they didn't like us. Ascend was a little too brash, and Promptus played the "underdog" role wonderfully. Picturetel didn't want Ascend to control video networking.

Ascend had won all of the other video manufacturers, so why were we worried? Well, Picturetel was the video sales leader at the time, and Ascend had built its marketing lead on a "follow the leader" mentality. That began to work against us. AT&T started to get cold feet, and we lost Telabs, an important telecommunications account. Telabs told us that they "wanted to go with a winner," Promptus.

Promptus gloated, giving our marketing guy a recruiting call. The implication was that the battle was over, Ascend had lost, and he should jump to the winning side.

I knew that we had to counterattack, but where? While Promptus was celebrating, we quietly investigated how Picturetel sold their equipment, which is how they would resell Promptus's product. We learned that 30 percent of their sales were through a sales force and that little mom-and-pop resellers, called value added resellers (VARs), sold 70 percent.

So we signed up the Picturetel VARs. All of them. We did it within weeks and announced nothing. In fact, we were spending time with our existing accounts, hoping they wouldn't jump ship. To Promptus it looked as if we were on defense, on the run. But we were on offense. The VARs we signed became a great sales channel for Ascend, and Picturetel ended up selling next to nothing for Promptus. The Picturetel sales force just didn't make enough money selling Promptus to make it worth their while to fight their own VARs. Speed and surprise made our counterattack work. If we had taken a long time to move, both Picturetel and Promptus would have shut down our tactic.

BUILDING BETTER PRODUCT:
THE BUCKET PLANNING PROCESS

The strongest defense is offense. Another way to lead the market is to produce quality products rapidly and continuously. The competition just can't keep up.

As the CEO, you should be driving development. The best way I've found to do this is to immerse yourself in your company's field of expertise. Learn your company's limits firsthand. Do your own probing for where the next ideas will come from.

> **TIP**
>
> I learned at Ascend to visit my engineering team regularly at the end of the day, chatting with whoever was hanging around. Becoming familiar with the products and the people who build them gave me an increased understanding of the new directions in which they, and my company, could grow. And I began to sense when schedules were on time, off time, or completely asinine.

Every company faces the issue of what process to use for finding the Next Big Thing. As you would expect, my method relies heavily on using chaos. I don't want an over-controlled environment. That may sound scary when you're talking about committing large resources, but my point is that you are not committing large resources blindly. You commit incremental amounts of resources based on measurable results.

At Entrepreneur America we develop new products using the "bucket planning process." We'd brainstorm on ideas for new functions for existing products, new products for new markets or existing markets, and variations of existing products. The ideas go into one of three buckets: Copper, Gold, and Platinum.

All ideas start off in the Copper bucket. The filter in this bucket is the company business model. Does the idea fit into the company sunflower? Does it make sense as a project under your business plan? If not, it stays in the Copper bucket. These ideas never get scheduled; no further action is taken on them. Ideas that do pass through that first filter go into the next bucket.

The Gold bucket is the staging bucket, where a few preliminary resources are allocated for pursuing the ideas. The filter here is customer interest. If there's no customer interest yet, then the idea stays in the Gold bucket. It gets no resources or schedules, but it does get further action. Someone is assigned to pursue the idea and talk to customers about it, gauging their interest level in the feature or new product. Gold ideas that customers have already asked about get upgraded to the next level.

Once in the Platinum stage, ideas must then earn customer commitment—a conditional purchase order. The conditions can be delivery date, functions, and/or price. Each idea also needs a little homework to demonstrate that the market for the product is larger than that single customer. The ideas that pass through those filters get scheduled and given completion deadlines with full organizational focus and resources.

Manage and execute this process for one year. You will end up leading your market, followed by the boards of discouraged, weaker players.

BE A SALES CHANNEL HOG

Here's a simple exercise that shows how important it is to lock up the key channels in your market. Write down all the channels of delivery for your product. Identify the top ten and apply the 80/20 rule. That means list the 20 percent that do 80 percent of the business. If you sign those accounts, you get 80 percent of your market by doing 20 percent of the work.

Here's an example from one of my Wanna-bes. Virtual Ink makes hardware and software for whiteboards that convert the Magic Marker scribbles into digital data. What the company should have done to establish market dominance was create a skunk works for their second product before the first one came out, ensuring that they would always be first to market with leading edge products. But they didn't have enough engineers on staff to focus on that.

Virtual Ink needed to do something to establish dominance, and quickly, because a competitor was breathing down their necks. I sat down with them and suggested taking the channel approach instead. We started by analyzing how whiteboards get out into the universe. A company called Quartet makes about 60 percent of the whiteboards sold, and a second makes another 10-20 percent. Clearly, if Virtual Ink locked up those two, it would pretty much control the distribution channels for its whiteboard add-on.

Targeting the top 20 percent of your market's companies is the best way to leverage your time. You're going to spend X amount of time selling, why not sell to the accounts that do 80 percent of the business? Get the names of the top five or so industry leaders for your product and go after them.

Virtual Ink landed Quartet and its biggest competitor. The CEO, Greg McHale, decided that locking up channels was so important that he spent six solid months on the road, banging on doors. He made being a start-up work for him. Most companies are busy selling more than one product, but Virtual Ink wasn't. Most CEOs are too busy to focus on big sales for six months, but this account was Virtual Ink's lifeblood. Greg had one thing to do—knock on doors and close deals.

Once he signed Quartet, he moved to Quartet's retailers. Companies like Steelcase, 3M, and Staples can source their whiteboards from anyone and might even think of developing their own branded similar hardware/software product. So Greg pounded on their doors and signed them up. He just kept moving down the food chain.

You can see where that leaves Virtual Ink's competitors. They're locked out. Everywhere they turn, there's Virtual Ink. What's left? Selling directly to individuals and corporations, which is an expensive and time-consuming way to do business.

It's not rocket science to develop the hardware that transforms whiteboard scribbles into computer data. Others will be able to replicate it, and Virtual Ink has maybe a six- to nine-month lead time. It's not enough, in terms of product development, but what really saves them is the fact that their competition has to knock on manufacturers' doors.

SMOTHER CUSTOMERS WITH SERVICE

Leveraging customer requests into power not only makes you a market leader, it allows you to command higher prices than the competition. If you become legendary in the way you handle customer needs, they'll pay more to use your equipment. Basically they wouldn't even dream of defecting to a competitor.

This kind of loyalty tends to grow through word of mouth. Once the ball is rolling, the company doesn't have to work hard to promote the image, just keep customers happy.

At Ascend we provided spectacular customer service at no charge. We didn't get into arguments with customers about whose fault it was, whether our equipment or someone else's software was causing the problem. We simply went out and fixed it. What we discovered was that although the problem often was not ours, solving it won us strong loyalty.

In the early days we had only two people in customer support, and we still managed to offer superior service. We used our engineers. When customers called with trouble, we put engineers on the phone and sent them to the customer. The customers were so impressed that we would send out the guys who actually built the hardware that they started telling their friends, and word of mouth won sales for us.

If you know you're going after the customer service angle of sucking the air out of the room, focus on that in your product. Build customer service right into the thing. That's what we did. Because of how we designed our hardware, Ascend's engineers could dial in remotely (with customer permission) and troubleshoot the problems from hundreds of miles away. Most of the time we could easily fix the problems that way, saving the customer time and hassle. Any time you do that, you win customer loyalty and word of mouth.

THE POWER OF THE MOVING TARGET

Moving targets are hard to hit. But developing new products and distribution channels isn't mobile enough. I tell my Wanna-bes to develop one entirely new *market* per year.

This is easy to say, very hard to do. I'm not talking about augmenting your existing market, adding new product features, or even making your product smaller/bigger/faster. Your existing products and services have markets. What I want to see is a company leveraging its core competency, developing new applications for it, and aiming it at a completely new market. Go back to your sunflower and start building petals.

Force yourself and your organization to open one new *market* per year, where stats are not immediately available for return on investment, size, rate of growth, or any other financial measure.

Begin working with *less* data, relying more on intuition, observation, and reasoning. You'll be able to work more quickly and hone your instincts as you anticipate your competition and execute your plans.

ANTICIPATION

Relying on instinct will help develop one of the most important tools for staying ahead of the market pack: anticipation. Until you learn to anticipate changes in the market and the competition, your company will just be a Wanna-Be has-been, dragging along behind smarter and faster start-ups.

There are two rules for anticipating changes in the market. First, when introducing a new product or service, make sure there is something that will drive its growth. Will it solve a pressing business problem? If not, there's no reason for anyone to buy it. The driver doesn't have to be something big. In fact, it probably won't be, especially if you're first in the field. But it should be a real problem that you're solving. If you have that, then sales will grow.

Here's an example of setting up drivers. I've just started working with a start-up called Lumber Tradex, based here in Montana. The company wants to create a business-to-business Web site for the lumber trading business. To drive the growth of the company, it has to link its product to solving some sort of problem for its customers.

The founders came out to Entrepreneur America, where we isolated three problem areas for lumber sawmills. First, marketing. These companies just don't market well or often enough to their customers. Second, accounts receivable. All of them have problems with aging receivables. Third, transportation. The logistics of moving lumber from one place to another are tricky and expensive.

So how can Lumber Tradex embed its product in one of these areas, solving the problem? If it can do that, the company will have a built-in driver to ensure that new customers pick up the product.

I sent the founders home with a research job: Go visit lumber companies and make a short pitch. Show them what kind of Web site you're going to build, then ask where Lumber Tradex can help their operations. Find out from the client if the need is stronger in marketing, accounts receivable, or transportation. And how much they're willing to pay. Presto—instant market research.

In Ascend's early days, when we were focusing on the undeveloped ISDN market, we didn't have anything driving our product. What we had was ISDN—a technology and a *hope* that it would be used for certain applications. It wasn't enough, and we had to bail on that market. The lesson was that there's a fine line between "anticipation" and "hoping."

The second rule of anticipation is to expect big changes but react to them with small practices. For example, Priceline.com anticipated a major change in how people buy and sell certain goods. They saw that manufacturing and service providers often have excess inventory—maybe a seat on a plane, last fall's fashions—that is hard to move. Why not bring together buyer and seller to negotiate?

Priceline.com started out small, selling just airline tickets. Now they sell almost anything, but starting small was smart. It was a low-risk way to test the radical new theory while building a name and earning some money to leverage themselves into other areas.

While you're watching the market, keep an eye on the competition also. By anticipating what your competitor is going to do, you can leverage your resources better and strengthen your weaknesses.

Anticipation can help you pinpoint your competitors' weaknesses. I'm not talking about making lists of your strengths and weaknesses versus the other guy's. That's completely the wrong approach. Unless your competitors are dodos, they know where their obvious weaknesses are, and they're in the process of correcting them. So their big failing, whatever you're patting yourself on the back about, will be neutralized before you know it. A better exercise is to find the *weakness in the other guy's strength*.

It's a Judo kind of thing, using your competitor's strength against him. Attacking the strength head-on isn't going to get you anywhere, but leveraging a little weakness can win the battle. Remember how David slew Goliath, chipping away with his little rocks at the weak spots until the giant fell.

I put this into practice when Ascend was competing against Cisco. Cisco's computer routers were the biggest, best, and most expensive a company could buy. That's because they could work with any routing protocol anywhere in any business. Well, what's the *weakness* inherent in that strength? All that routing capability was expensive, and not every customer needed every protocol. In fact, Internet access used only a single protocol—TCP/IP. Ascend began building competing routers with just TCP/IP protocol that cost less than half the price of Cisco's. That strategy succeeded. Trying to build the same router Cisco built would have put us out of business.

Be a moving target—keep changing product, making it impossible for competitors to match it. Selling against your company will be a nightmare.

EXERCISE:

Sales Questions

1. What are your key sales channels?

2. What are the most highly leveraged channels?

3. Who within each of those channels are the key accounts?

4. Who within those accounts controls the decision about using your product?

5. Why do they need your product?

6. What does it do for them?

7. What critical subset of the key channel accounts would give your competitor a nightmare?

Market Questions

1. Who are the key influencers in the press?

2. Who are the key analysts?

3. What key alliance would swing momentum your way?

Engineering Questions

1. What technology breakthrough would change the playing field?

▶ You are welcome to refer to the Entrepreneur America Web site for additional information (www.entrepreneur-america.com).

 *So You've Got the Money,
Now What?*

ROAD MAP: ▶ Congratulations, you've signed your first investment deal! Now the hard work really begins. It's time to make the transition from start-up to professionally managed company. Here's how to beef up your team and handle the board.

Imagine depositing your first big VC check. Millions of dollars at last! What will you do with it? Buy a car? Lease an office building? If you're like my successful start-ups, you'll probably write yourself your first paycheck. And it won't be big, either.

Virtmed's founders worked for six months without seeing a penny. When they won their first round of capital, they finally started paying themselves a whopping $60,000 salary. Same for the founders of Actuality. Salary tends to be a big event for my entrepreneurs because most haven't seen income in a long time. They've been bootstrapping their baby company (and paying rent) off Mom's generosity or angel money.

Start-ups are all fixated on raising capital, but they never really consider what, specifically, they'd do with a $10 million check. How do you control spending, and what do you spend money on? Take my advice and set up a spending plan now, *before* you get the money.

Fill the Critical Positions

Once you do get a little cash, the first thing I advise is assessing what critical positions need to be filled, then spending the money to fill them right away. Most high-tech start-ups need a VP of engineering and several engineers. A VP of marketing and VP of sales come in handy, as does a CFO.

Of course, you're not going to be able to hire them all at once. I know from experience how difficult it can be for start-ups to recruit first-rate talent. It can start to feel like a catch-22 situation: you can't find great management because you've run out of money, yet you can't get more money until you hire great management. The problem is that there's a high correlation between business success and filling key positions quickly with good people, so you've got to have them.

When you're in bootstrapping mode, the thing to do is stagger the people and the money. Your co-founding team members, like you, were probably willing to work for tiny salaries (or none at all). Seed capital will buy you a few key employees. The strengths they bring to the company will help win more investment, which will buy more employees, and so on. At a certain point, of course, revenues will kick in.

But once the VCs hand over a $3 million to $5 million check, they're going to expect you to spend it to fill any empty slots. Make a list of the crucial people in management, engineering, marketing, sales, finance, and operations. Select one person to be accountable for hires in each department (there might be some overlap if you're still seriously understaffed). Normally it's the director or vice president—absent them, it's you. This person should report at each board meeting on the status of hiring in his or her department.

At Ascend we were always hiring. It's so difficult to find "A"-level people that we just never stopped looking.

FINDING THE TALENT

My best luck in finding people has always been networking. Use your contacts and the board members' also. They'll have better Rolodexes than you do.

But I have to admit that for major hiring decisions, headhunters are becoming a necessary evil. I say "evil" because in my experience it's hit or miss finding good candidates this way. Sometimes headhunters work, sometimes they don't. But since fast-growth companies don't always have time (or the in-house manpower) to conduct a massive search, more and more are dependent on headhunters. If you're going to use them, ask your board members for recommendations and make a short list of those with strong track records (more than one placement) in your industry *and* for the position you're filling.

Interview the firms, focusing on the process they use to find candidates. How rigorous is it in terms of tracking, sourcing candidates, and interviewing them? The successful firms have strong processes and good people implementing them. Weak firms end up with hit-or-miss results. I tell my Wanna-bes that going with a large headhunter doesn't necessarily mean you get a good value for the fee you pay. The service is only as good as the individual assigned to your search.

If the firm isn't producing candidates after a couple of months, don't let the situation malinger. Cut your losses and fire the firm. Sure, you'll be out some up-front money, but finding good candidates quickly is a higher priority.

WAITING FOR THE PERFECT CEO

If your company is one that needs a CEO (and most are), by then you'll have investors and a board to help with the search and interviews. A good candidate will want to meet the entire team, and they'll all want to interview him or her. This is like a marriage—hire in haste, repent in leisure! Boards won't fire CEO candidates lightly, so once you've hired one, you're pretty much stuck with that decision.

At Silicon Spice, a start-up I've worked with closely, the board and I insisted from the start that the company recruit world-class talent from the semiconductor industry. That caused a lot of anxiety because it started looking as though we'd never settle on the right person.

The company, which was designing and building a new semiconductor chip, needed a top-notch CEO. The young founder, Ian Eslick, wisely realized that although he knew the technology, he didn't have the experience to run the company and inspire confidence in investors and customers. A recent graduate, he hadn't even worked at a business yet, much less run one.

Unfortunately for the California-based start-up, Silicon Valley suffers from a deficit of good CEOs. The place is awash in high-tech start-ups, all of which desperately need a seasoned leader. So there's huge competition for someone with the right qualifications. We interviewed dozens of candidates, but none seemed good enough. At times Silicon Spice's board, which included some really big-name VCs, got nervous because the search was taking too long. I was acting CEO, and even though I had a lot of anxiety about the search, I fought back.

"If you want to be angry at me, go ahead," I announced at one of the board meetings. "But I'm not going to bring in someone who isn't a world-class kind of guy. It would be really easy to get you all off my back by hiring one of these other Wanna-bes. But in about six months you'd be desperate to get rid of the guy, which is a nasty position to be in."

Dumping a CEO or a board member—any key management hire—costs the company a lot of time and money. It also sends morale right down the tubes. Fortunately the board eventually agreed to take the time to do it right.

The wait paid off. In spring of 1998 Silicon Spice landed a great catch—Vinod Dham, the father of the Pentium processor, who had led Intel's microprocessors group. Vinod had been working on a similar business plan of his own and happily joined Silicon Spice so that he didn't have to start his idea from scratch. He instantly helped raise $30 million from Cisco, IBM, and other top investors—enough money to create a team of sixty-five first-class engineers. Today the company has about a hundred employees and has raised about $70 million in venture capital.

TITLES: WHO'S ON FIRST?

As you start setting up your management structure, be careful of the CEO/COO thing. It's not a good structure to put in place during the start-up phase.

A start-up must have a leader, a visionary, a decision maker. That person is the driving force. If that role is divided between a chief executive officer and a chief operating officer, there are three possible results. First, the COO is a weak, ineffective drone and therefore useless. Second, the COO is really strong, intelligent, and ambitious, creating power struggles. The third option is the only good one. The COO is strong and ambitious but tempered by a lifelong friendship with the CEO. They can comfortably take advantage of each other's strengths and shore up weaknesses. This is a great configuration, but the odds aren't good for getting it.

At Ascend we used the CEO/COO configuration, but only after we went public. The main reason was that I was about to have back surgery, and we needed to outline a clear chain of command for the prospectus. The COO had been my top manager for many years and did his job so well that he ended up running Ascend after I left.

ORGANIZATIONAL CHART—1

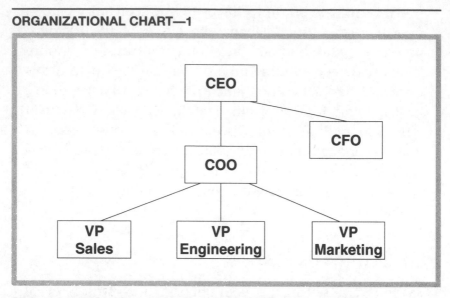

Slide 1: I don't recommend this configuration. It's unstable and can set up power struggles or undercut the power of the CEO.

ORGANIZATIONAL CHART—2

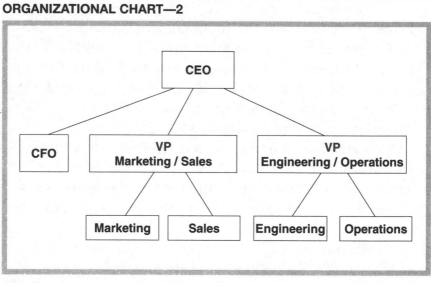

Slide 2

Manage to the Measurables

Hiring and setting goals—those are the two main operations in a small company's life. They're inextricably linked

ORGANIZATIONAL CHART—3

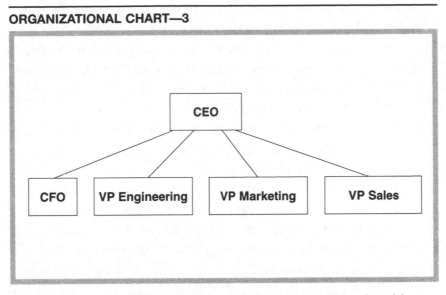

Slide 3: Any permutation of the structures shown in slides 2 and 3 provides a good, balanced distribution of power and control.

because you can't start hiring until you have goals in place, otherwise you won't accurately project your staffing needs. Each goal should have certain "performance triggers" that signal changes in funding for different parts of the company. This is always a big epiphany for my Wanna-bes: working like a performance-oriented company with measurable goals.

Most budgets are "time triggered." For example, a typical budget would say, "Add two engineers by June." Fine, but what if you don't meet your revenue goals, or what if demand for the product is low? More engineers might just confuse matters further, plus you'll go broke payrolling them.

A performance trigger instead says, "Add one engineer after X event is achieved." Notice that I added only one engineer. With performance triggers you get only one hire at a time. That's not only because I'm sort of old-fashioned about not hiring until the income stream merits it, but also because it's so difficult to find good talent. Even if you list two or three hires, they rarely come in as a group. In the real world you're going to find only one at a time.

Performance dictates head count. In most companies teams always meet head counts but frequently fall short on performance. But peg them together and head count will never outpace performance.

Say you missed revenue targets. During the quarter you should have enough information to realize that things are not shaping up properly. So start cutting back *during* the quarter. Not drastically, just a gentle throttling back by delaying some hires, deferring others. The idea is to moderate the head count dollar burn in small increments of deferrals and push-backs at the early signs of a bad quarter. That's much smarter than blindly throwing labor dollars at your company with no measure of performance.

SETTING GOALS FROM THE BOTTOM UP

The structure that works the best is when the CEO pushes the responsibility down to the department heads, who should push it further down to the employees. At each level critical information gets contributed. For example, say the CEO decides that her networking company is going to be the market leader in small-business computer network installations.

Then the CEO pushes that goal down to the managers, who decide how to measure it. Marketing might suggest tracking new program roll-outs each year and mentions in trade publications. Engineering may gauge new systems that the team learns to install. Sales might count word-of-mouth sales.

Front-line employees help finalize the goals by determining what the numbers are. How many new marketing programs should be introduced each quarter? Each year? How many word-of-mouth sales do the salespeople have to make?

There are two payoffs to this approach. First, the company has a highly accurate gauge for measuring success. Nobody knows the specific numbers, and the potential for reaching higher, better than the individuals doing the jobs. The second big benefit is that by pushing goal setting down the line, it ripples up and down the company, becoming part of the culture. Everyone knows that the company runs on the numbers, and it's clear to all when things begin to slip.

THE ANTIBUDGET

I never get too obsessed by budgets. Instead I let performance triggers dictate how much to spend on departmental projects.

The key is to insist that managers perform first, and only then do their projects get financed. Let's say a manager wants three people for some task. I like to ask, "Would you throw in the towel if you could get only one?" If the answer is "no," then start the manager off with one or two people and monitor the progress. Watch what happens and invest more resources as the project develops. If the answer is "yes," then it's probably a worthless project anyway.

One thing to bear in mind when setting up budgets and performance standards is that bad quarters are defined not just by revenue and/or earnings. In fact, start-ups might not yet have either of those. Performance is measured by misses—are you falling way behind on production schedules? This is a classic start-up problem. The typical response is to hire more employees, blaming the problem on a lack of bodies. My response is to study what is causing the slips.

Probably the employees have too much to do in too little time. You could adjust the time frame, but then you fall into the trap of always pushing back the release date. You could hire more bodies, but that ends up being an expense that rarely pays off. The best thing to do is adjust the functions/features within the product so that it fits within the given time frame. It's hard to get five pounds of dirt into a two-pound bag. Process the features down to the minimum that the customer needs to get started. You can always add more later, and it gives you an excuse to stay in touch with the client.

Remember that budget head counts are more than just body counts. Head count is a description of employees' seniority, experience, pay, and role in the company. Being specific can help avoid a classic mistake: overbudgeting on the executive side. I've seen lots of companies create projections that are top-heavy with VPs and directors, and that's because they forget just how expensive they are. A detailed head count differentiates between the executive who will have to be offered $200,000 plus stock options and the $35,000 office assistant. Also, it takes into consideration the time and resources needed to hire each.

Boards: The Company Police Force

If you have investors (who sit on the board), then you've cleared the first hurdle of becoming a successful company. Naturally those investors want to keep a close eye on what you're doing with their money. Smart companies use that watchful eye to their advantage—after all, those investors aren't stupid, and they'll have good advice.

In the early stages of a company I recommend holding eight to ten board meetings per year. These experienced people are supposed to be your advisers; talking to them often can only help your struggling start-up. As the company

▼ **TIP**

Some companies develop two stories, "good news" for the board and the truth for everyone else. That's ridiculous. If you truly want helpful advice from your board, stick with one story, whether it's good or bad.

matures and starts making money, drop to six a year; you'll find that you need less frequent supervision than when it was young.

All the board members should attend each meeting. Ideally, all should be physically present, but typically one or two busy travelers are phone-conferenced in. I'm on fourteen boards, so it's just not possible for me to make all meetings in person. Do yourself and your board a favor—invest in a good conferencing system with multiple microphones.

CHOOSING BOARD MEMBERS

The company CEO will sit on the board, as will one representative from each investing firm. If you've had early seed capital, you might also have an angel on your board. So, if there are two lead investors in your first round of VC money, that's two board members. (In later rounds investors don't usually bother to sit in on an already solid board.)

If you want to keep the board to a manageable size (about five to seven people), that leaves about two slots to fill. Look for individuals with track records of success in your area, entrepreneurial people with operating experience. Don't put your friends and family members on the board unless they have special business skills. It might be tempting to build up a board that will "rubber stamp" every choice you make, but you won't learn anything that way. Nobody will question your bad decisions or make insightful suggestions.

It can be tricky deciding who from the company will sit on the board, especially if you have several co-founders. My advice is to keep it simple. One insider, that's it. Otherwise it tends to get awkward having people report to you while they're also sitting on the board. Remember, the board is basically *your* boss, and anybody attending meetings has free access to board members. It can get pretty strange.

There are, of course, exceptions to my rule. If your company has a CEO/COO configuration (and it's cemented through some kind of lifelong friendship thing), then both sitting on the board is fine. When companies go public, the CFO is added to the board.

KEEP THE BOARD FROM GETTING BORED

Let me give an example of what happens all too often in board meetings. The CEO walks in, greets everyone, and launches into a summary of accomplishments, including slides. Ninety minutes later we're still bogged down on the initial two slides, while the rest of the company management team taps fingers and rustles papers, waiting to make their presentations. But with such a long-winded start, the CEO has stolen most of the good material. The executives, fighting for their moments in the spotlight, end up repeating most of the data already covered. Two or three of the executives get cut from the program once it starts to run over three hours. Board members begin quietly to slip out of the room. It's a mess.

The CEO's job is to open the meeting, keep it moving, approve the minutes from the previous meeting, then *listen*. You've signed these board members on to help guide and advise you, so let them do their job. If you start off with a summary of accomplishments and defeats, your team will never get the chance to present their story. Sit down and let your team do the work.

A two-hour meeting (or less) is perfect. Face it, people get bored, and they're also busy. If you extend much past two hours, members will begin walking out anyway. Go over that time limit only when major circumstances dictate.

The strangest board meeting I've sat through was when the CEO invited some consultants to make a presentation. The right way to do it would have been to warn the board members in advance and then bring in the consultants at

**Minutes of a Meeting
of the Board of Directors**

February 18, 2000

A telephonic meeting of the board of directors (the "Company") was held on Friday, February 18, 2000. Participating were directors: Greg G, J.H., S.C., and R.R. Also present at the invitation of the Board were Greg A and K.S. Ms. C acted as secretary for the meeting.

Mr. Greg G, the Chairman, called the meeting to order at approximately 5:15 P.M.

Mr. Greg G and Ms. C provided an overview of recent discussions they had had with various investment banking firms regarding the Company's potential initial public offering. There was a general discussion regarding the pros and cons of the various investment banks, including their corporate finance and research capabilities. After discussion, the board determined that Credit Suisse First Boston should be selected as the Company's lead manager and the Company should include Thomas Weisel Partners, D.R. and A.H. as co-managers, and others as Mr. Greg G and Ms. C, after consultation with CSFB, deem appropriate. Mr. Greg G and Ms. C were directed to inform CSFB and the other potential co-managers of the Company's decision.

There being no further business to come before the meeting, it was, upon motion duly made and seconded, adjourned at approximately 6:15 P.M.

_____
Ms. C, Secretary

Approved:

_____
Mr Greg G, Chairman

Slide 4

the end of the meeting. But the CEO sprang them on us at the beginning, inviting the consultants to sit through the entire meeting. So much for "no surprises!" The consultants dragged out their slides, and three and a half hours later we were all still there, bored stiff. After the meeting I let the CEO have it. Fortunately he hasn't made that mistake again.

One of my Guts and Brains Wanna-bes who really knows how to run a board meeting is Greg Gianforte of RightNow Technologies. He walks in and stands up for two or three minutes settling administrative matters, stock issues, and so on.

Minutes of the Previous Meeting (Slide 4). Next, the board approves the minutes of the previous meeting. Note how brief these minutes are. There's no reason for them to stretch longer than a page. I've seen minutes that are four pages long, listing the minutiae of the meeting, like "The board discussed the XYZ product and why its schedule slipped for the twelfth time." That level of detail is a liability. In a lawsuit the board meeting minutes get deposed, so the briefer they are, the better. Rather, say something like "The board reviewed engineering."

Once the board approves the minutes, Greg stands up and shows three slides.

The Agenda (Slide 5). The "agenda" slide is fairly self-explanatory. It spells out which managers will be giving departmental status updates.

Company Status and Goals (Slide 6). The "status" slide gives a brief summary of action items from the last meeting and what's to come in this one. For each measurable area there is a goal, the actual progress, and what percentage that progress is of the goal. "Major initiatives" focuses generally on company goals with time frames of ninety days or less. "Future initiatives" are longer-term goals. "Lows" and "highs" are where the executive team grades itself on goals accomplished or missed since the last meet-

Meeting of the Board of Directors
April 4, 2000
11:00 A.M. to 4:00 P.M. (Mountain Time)

Roaring Lion Ranch
77 Storm King Road, Hamilton, Montana 59840
Phone: 406.363.0164
Fax: 406.363.0155

AGENDA

I.	**Call to Order**	Chairman Greg G
II.	**Minutes for Ratification**	
	Board of Directors Meeting, February 18, 2000	
III.	**Operations Review**	
	A. CEO's Report	Greg G
	B. President's Report	J.H.
	(1) Sales	
	(2) Customer Support	
	C. Marketing and Business Development Report	J.P.
	D. Development Report	M.M.
	E. Finance Report	S.C.
IV.	**Legal Update**	J.M.
V.	**Detail Reviews**	
	A. Product Financing/Accounts Receivable	S.C.
	B. Compensation Review	Greg G
	C. Ernst & Young Management Letter	S.C.
	D. IPO Update	Greg G
VI.	**New Business**	
	A. New Board Members	Greg G
	B. Options Granted/Pricing	S.C.

Slide 5

ing. "Action items" is an important section—good companies never lose track of them.

The organization slide lists employee names, titles, and
reporting responsibility for the whole company.

Company Org Chart (Slide 7). This one is important because young companies like this change rapidly. What the
board wants to know is how well the company is progress-
ing toward filling key slots, particularly managerial.

Greg skims through his presentation because his man-

Company—Status and Goals (4/3/2000)

Status

	4Q99	Q1 Goal	Jan	Feb	Mar	1Q00	%	Q2
Bookings	$3.0M	$5M	$1.5M	$2.0M	$2.5M	$6.0M	120%	$7.0M
New Customers	100	150	20	40	60	140	93%	300
Total Customers	500	800	600	650	700	750	94%	850
Avg. Deal Size	$60K	$65K	$40K	$45K	$50K	$50K	77%	$70K
Head Count	150	250	150	190	230	250	100%	270
DSO	50	70	80	80	90	70	100%	80
Receivables	$2.0M	$3.5M	$2.0M	$2.5M	$3.0M	$3.5M	100%	$5.5M

Major Company Initiatives

1. Hit sales goal of $6M—J.H.
2. Exec. hires—done except VP customer service
3. ASP hosting. 70% of new customers hosted, world-class facility—B.F.
4. Development staff of 45 by end of quarter—M.M.
5. 95% of customers live within customer expectations. Customer satisfaction of 3.7—B.F.
6. RNW3.2 shipped, RNM1.0 shipped—M.M.
7. Develop corporate marketing plan. Establish Bay Area office—J.P.

Future Company Initiatives

- Acquire outbound marketing product
- Europe sales
- Asia sales

Slide 6

Highs

- Exceeded quarter bookings goal
- 150 customers live, log jam cleared
- Organizational meeting
- New Web site
- New building
- Board members

Lows

- Tax Issues
- Hosting %

Board Action Items

	Date	Person	Description	Status
1.	07/21/99	B.F.	Customer support must be self-sufficient and not dependent on Mike for technical assistance	DONE
2.	1/10/00	All	Aggressively hire all exec. positions	Only VP customer service remaining
3.	3/8/00		Metrics for getting people live	Goal updated—DONE
4.	3/8/00		Likely future of slipping terms on deals	Update at meeting
5.	3/8/00		Analysis from Dorsey on option grant pricing	DONE

Slide 6 (continued)

CORPORATE STRUCTURE

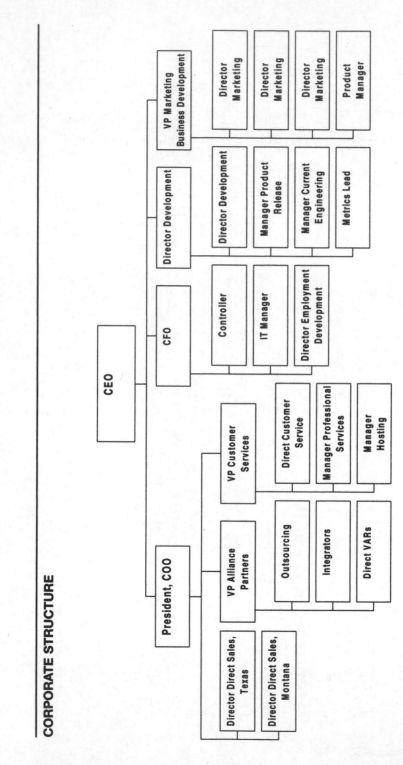

Slide 7

agers will give more detailed information. He sits down, and for the main part of the meeting he invites his managers in and lets them run the brief slide show that gives updates on the company's progress.

The departmental heads (sales, marketing, engineering, operations, and so forth) take turns presenting their material. Each department takes fifteen to twenty minutes to show two slides—a status slide and a head count slide. The six-month goals have been worked out in advance within the company, and they don't change from meeting to meeting. Only the progress does. Each goal is measurable with hard numbers.

Greg's managers use objective numbers to show progress on each of their status/goals slides.

Sales Status and Goals (Slide 8). The slide from the sales department shows "bookings" and "new customers," which are great measures of how well the product or service is being received. "Average deal size" indicates how much the company makes on each customer. "Total pipeline" measures what the company would earn if all the deals in the pipeline closed. "Telesales" metrics relate to that department's productivity. These same metrics can be applied to other sales configurations.

Sales Org Chart (Slide 9). This sales organization chart goes into more detail on the lower levels than the management chart. The detailed head count illustrates where the company is staffing up and where it needs to fill positions.

The "metrics" section of this chart focuses on the key measurable goal for customer support. The format is the same as the other departmental status slides—show what you plan to do (goal) and what you've done (actual) by month and quarter. "Low" and "high" focus on things that went well or badly since the last board meeting.

Customer Support Status and Goals and Org Chart (Slides 10 and 11). The customer support status has the same format as the sales section but focuses on customer support issues.

Sales—Status and Goals (4/4/00)

Status

Slide 8

		Q4	Q1 Goal	Jan	Feb	Mar	Q1	%	Q2 Goal
Bookings		$ 3,000,000	$ 5,000,000	$ 1,500,000	$ 2,000,000	$ 2,500,000	$ 6,000,000	120%	$ 7,000,000
Bookings Detail									
	MT Telesales	$ 2,000,000	$ 4,000,000	$ 1,000,000	$ 1,000,000	$ 1,000,000	$ 3,000,000	75%	$ 2,160,000
	TX Telesales	$ 500,000	$ 500,000	$ 250,000	$ 250,000	$ 1,000,000	$ 1,500,000	300%	$ 3,240,000
	International	$ 250,000	$ 250,000	$ 200,000	$ 250,000	$ 250,000	$ 700,000	280%	$ 300,000
	OEM	$ 250,000	$ 250,000	$ 50,000	$ 500,000	$ 250,000	$ 800,000	320%	$ 300,000
Total		$ 3,000,000	$ 5,000,000	$ 1,500,000	$ 2,000,000	$ 2,500,000	$ 6,000,000	120%	$ 6,000,000
New customers		100	150	20	40	60	140		300
Avg. Deal Size		$ 60,000	$ 65,000	$ 40,000	$ 45,000	$ 50,000	$ 50,000	94%	$ 40,000
Avg. calls per rep/day		30	-	40	40	40	40		
Total Pipeline—5X			25M			28M			35M
March Pipeline—2X						5M			5M
Telesales Head Count Total Reps		60	$ 82	60	60				
0 to 30 days # reps		10		10	8	10	9		
Total Revenue		$ 60,000		$ 60,000	$ 60,000	$ 60,000	$ 60,000		
Productivity per rep		$ 6,000	$ 10,000	$ 6,000	$ 7,500	$ 6,000	$ 6,500		
31 to 60 days # reps		10		12	12	10	11		
Total Revenue		$ 40,000		$ 40,000	$ 40,000	$ 40,000	$ 40,000		
Productivity per rep		$ 4,000	$ 30,000	$ 3,333	$ 3,333	$ 4,000	$ 3,556		
61 to 90 days # reps		10		8	10	10	9		
Total Revenue		$ 100,000		$ 100,000	$ 100,000	$ 100,000	$ 100,000		
Productivity per rep		$ 10,000	$ 45,000	$ 12,500	$ 10,000	$ 10,000	$ 10,833		
91+ days # reps		25		25	25	25	25		
Total Revenue		$ 800,000		$ 800,000	$ 800,000	$ 800,000	$ 800,000		
Productivity per rep		$ 32,000	$ 60,000	$ 32,000	$ 32,000	$ 32,000	$ 32,000		

Initiatives

1. Meet productivity goals
   A. $125K run rate for tenured rep
   B. 40 calls per day
   C. Attain $5M in Q1

2. Hiring Goals
   A. 82 Telesales Reps by 3/31
   B. VP of International
   C. VP of Indirect

3. Organizational Structure
   A. Identified 82 territories
   B. Split country between MT and TX
   C. Implement sales structure

Status

March actual $106,440
March actual 37 calls
Q1 actual $5.0M

Actual 79
Done, 4/23/00 start date
Done

Done
Done
Done

Lows

• 31 people sold $0

Highs

• 4.7% of quarterly #
• 96% attainment of hiring goal
• 8 reps over $100K in March
• Centralized SFA package selected
• 31 alliances
• 10% of revenue in March incremental through VARs

Slide 8 (continued)

SALES—ORG.

Director of Sales

- **Manager West**
  - Person 1
  - Person 2
  - Person 3
  - Person... "n"
- **Manager East**
  - Person 1
  - Person 2
  - Person 3
  - Person... "n"
- **Sales Manager**
  - Person 1
  - Person 2
  - Person 3
  - Person... "n"
- **Sales Director, Customer Service**
  - Person 1
  - Person 2
  - Person 3
  - Person... "n"
- **Systems Engineer Manager**
  - Person 1
  - Person 2
  - Person 3
  - Person... "n"

Slide 9

Customer Support—Status and Goals (4/4/00)

Status

Metrics	Q4	Q1 Goal	Jan	Feb	Mar	Q1 Total
Live	100	200	30	40	90	160
Not live	150		150	110	85	80
Ready to go live				35	32	31
Not live >30 days	90		100	80	50	45
Rev. not live	$ 2,500,000		$ 2,700,000	$ 2,400,000	$ 2,000,000	$ 2,330,000
% customers live in 30 days	30%	99%	10%	25%	30%	29%
Customer satisfaction	3.5	3.6	3.7	3.7	3.8	3.75
% of new customers who host	30%	65%	60%	55%	55%	54%
% of total customers who host			50%	55%	55%	52%
Professional services rev.	$ 200	$ 555,555	$ 35,000	$ 41,000	$ 60,000	$ 136,000
Hosting Revenue			$ 80,000	$ 85,000	$ 166,679	$ 331,679

Initiatives

1. Meet Productivity Goals
   A. 80% live in 30 days
   B. 70% of new customers hosting

2. Meet Hiring Goals
   A. 28 customer service reps by 3/31
   B. Hire VP customer service

3. World-Class Hosting Facility
   A. Exodus
   B. Oracle DB2 admin.
   C. Retained search for manager

Status

March actual 29%
March actual 54%

Actual 30
4 candidates identified, decision by 4/30/00

Done
Done
2 candidates identified

Highs

- Full head count
- Customer satisfaction highest ever
- Reliability of hosted sites
- 91 people live in March, 154 in quarter

Lows

- Difficulty in hiring tier 2 customer service reps.

Slide 10 (continued)

CUSTOMER SUPPORT—ORG.

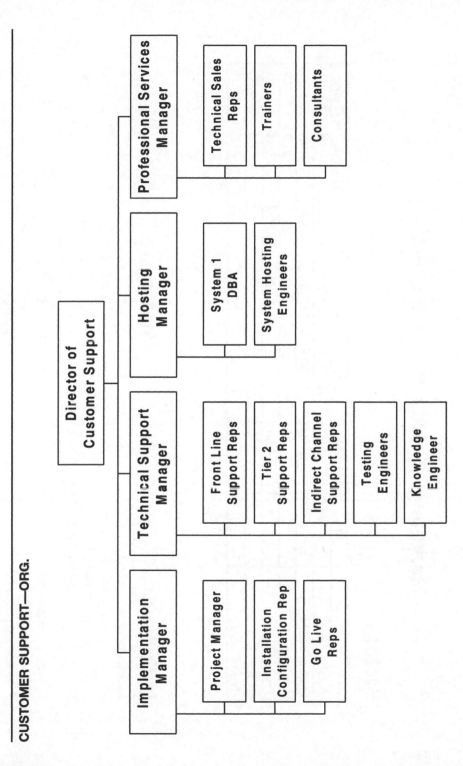

Marketing—Status and Goals (4/4/00)

Status

	Q4	Q1 Goal	Jan	Feb	Mar	Q1–2000	Q2 Goal
Total Leads	9,000	38,000	10,000	2,000	4,000	16,000	14,000
Online Leads	5,000	10,000	3,000	1,300	1,500	5,800	8,000
Seminars	0	150	75	250	200	525	850
Trade Shows	0	1,500	0	500	1,000	1,500	800
Other	90	950	75				
Head Count	10	15	15	15	17	17	18

Initiatives

1. Seminars
   A. Q1 seminars—7
   B. Q2 seminars
       3 planned per month

2. Web site redesign

3. Reference program

Status

Attendance 364

Done 3/31/00

Ongoing

Slide 12

4. Partner program — Ongoing
   A. Demos scheduled with NASDAQ and Olympics

5. Webinars—April 27—introduce 3.2 — Ongoing

6. Hiring — Ongoing search
   A. RNW product manager

7. Return on investment
   A. Press — Q1 $4.50 per $1.00 spent
   B. On-line Adv. — Q1 $3.50 per $1.00 spent
   C. Seminars — Q1 $6.00 per $1.00 spent
   D. Trade shows — Too early to tell
   E. User conference — Q1 $15.00 per $1.00 spent

Highs

- Seminar series
- 5 trade shows this quarter
- *PC Week* review/*Internet Week* review
- Redesigned Web site launched

Lows

- Web site cost exceed budget by 30%
- Quiet period impact on PR

Slide 12 (continued)

Marketing Status and Goals and Org Chart (Slide 12 and 13). Goals for the marketing department can be tactical, like RightNow's. The company focuses on leads, number of trade shows, and seminars. Or goals could be strategic, such as assistance in landing a certain number of specific accounts. The "initiatives" here highlights the key goals for the quarter.

The marketing head count slide gives an at-a-glace update on positions and openings.

Development Status and Goals and Org Chart (Slides 14 and 15). This slide, for the engineering department, is divided into "head count" and "development projects" sections. Head count lists hiring goals and accomplishments, including "pipeline," which indicates flow (number of good candidates). The second section lists projects and where they stand in relation to their planned release dates. From this slide, board members can instantly see what engineering phase each project is in.

For RightNow's board, the engineering head count slide is mission critical, so it's slightly more detailed than the others. It includes all titles and hiring dates for interns.

Finance and Administration Status and Goals and Org Chart (Slides 16 and 17). In these slides the "finance & admin" is a head count number. "F&A % of bookings" shows the percentage of revenues spent on this department. "DSO" refers to days sales outstanding, "net cash" is the company's available cash, and "new customers" reflects update of the products. The "initiatives" section is broken into different subdepartments, with their goal dates.

MARKETING—ORG.

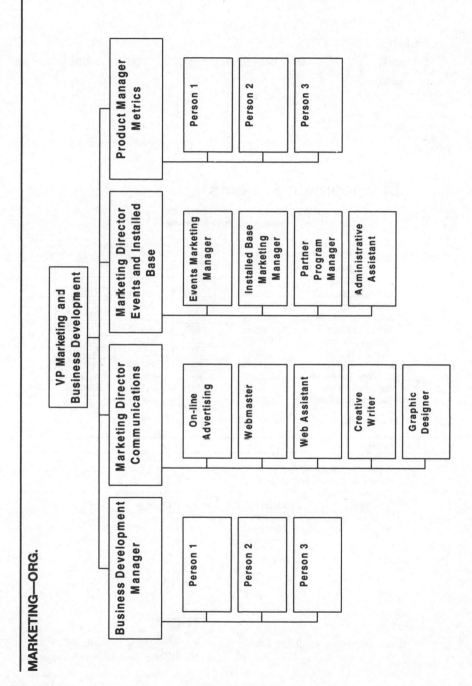

DEVELOPMENT—STATUS AND GOALS (4/4/00)

Status

Position	Q4	1Q00 Goal	Jan.	Feb.	Mar.	Q2
Director	2	1	1	1	1	
Project Lead/Mgr.	5	5	4	4	4	
Developer	15	0	18	18	20	
QA	4	5	3	3	4	
Language Coordinator	0	1	0	0	1	
Documentation	1	2	1	2	2	
Student Interns (.5 FTE)	1	5	1	4	4	
Total (FTE)	27.5	16.5	27.5	30	34	45

Development Projects

Development Project Tracking

Milestones	Metrics 1.0		RNW 3.2	
	Plan	Actual	Plan	Actual
Kickoff Meeting	10/25/99	10/29/99	1/17/00	1/21/00
UI Design Review	9/24/99	9/30/99	1/28/00	2/8/00
Project Plan Review	10/25/99	10/29/99	1/28/00	1/28/00
Design Review	9/24/99	9/24/99	2/2/00	2/3/00
Feature Complete Review	10/25/99	1/5/00	2/23/00	2/25/00
System Test Start	10/25/99	1/17/00	2/28/00	3/2/00
Platform Test Start	10/28/99	2/7/00	3/13/00	3/27/00
Internal Trial Start	11/1/99	3/2/00	3/27/00	3/29/00
Customer Trial Start	11/8/99	3/9/00	3/31/00	4/8/00
Customer Trial Review	11/22/99	4/3/00	4/12/00	4/20/00
Software Complete	11/26/99	4/7/00	4/17/00	4/24/00
Customer Availability	1/15/00	4/7/00	4/29/00	

Project Team

Lead	M.C.	W.E.
Staffing	2	13
Project Sponsor	T.K.	Product Mgr.
Release Coordinator	J.M.	J.M.
Documentation Developer	J.H.	J.H.
Customer Sponsors	RNT	Pumatech
	Myron	Montage
	Zap.com	
	Osmonics	
	LinuxCare	

Lows

- Space shortage/new building delay

Highs

- RNW 3.2 in internal trial
- RNW 3.1 available in 8 languages

Slide 14

DEVELOPMENT—ORG.

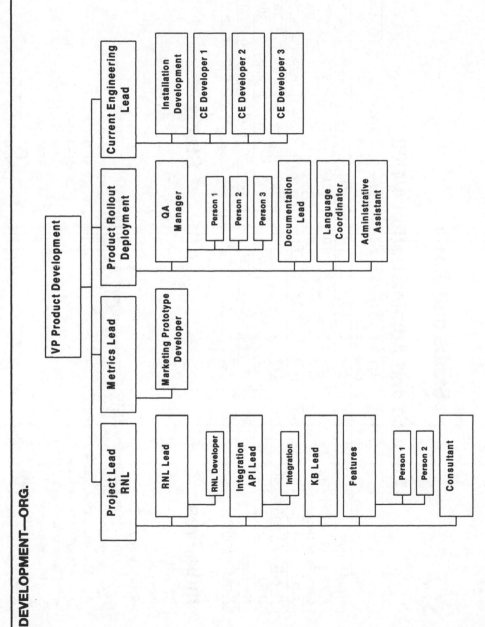

Status and Goals

Finance and Administration 4/4/00

	Q4	Q1 Goal	Jan	Feb	Mar	QTD	%	Q2 Goal
Head Count	150	250	150	190	230	250	100%	270
Finance and Admin.	15	20	20	20	21	21	95%	21
F&A % of Bookings	10%	10%	10%	10%		-	-	10%
A/R	$2.0M	$3.5M	2.0M	$2.5M	$3.0M	$3.5M	100%	$5.5M
% A/R > 60 old/past due	10%	11%	14%	12%	13%	13%		12%
DSO	50	70	80	80	90	70	100%	80
Net Cash	$15M	$12.0M	$14.0	$13.0	$12.0	$13.0M	102%	$18M
New Customers	100	150	20	40	60	140	93%	300

Initiatives

Status

Finance

1. Prepare for IPO — Ongoing
   (due diligence, drafting, ESPP, directed share program, D&O insurance, indemnifications, Delaware entity, draft MD&A, printer, transfer agent, road show)
2. Hire A/R, admin., project finance — April
3. Support move in dates — April/May
4. Clean, quick quarter close — April
5. UK subsidiary — April
6. Finalize HR policies — April
7. Q2 detail budget — April 15
8. Decrease %> 60 days—collections cesses — April

Slide 16

Information Systems

9. Implement front office solution — Ongoing
10. Support move in dates — April/May
11. Support world-class hosting initiative — Ongoing

Human Resources

12. Recruit fast and well—new director plans — Ongoing
13. Develop employee meetings/events — April
14. Employment agreements/compensation — Now

Legal

15. International reseller agreements — April
16. Indirect channel agreements — Ongoing

Highs

- Quarter results
- Organization meeting
- Management training
- HR pro

Lows

- Tax issues
- Forecast revision (marketing spend)
- Indirect support (legal and business)
- DSO

Slide 16 (continued)

FINANCE AND ADMINISTRATION—ORG.

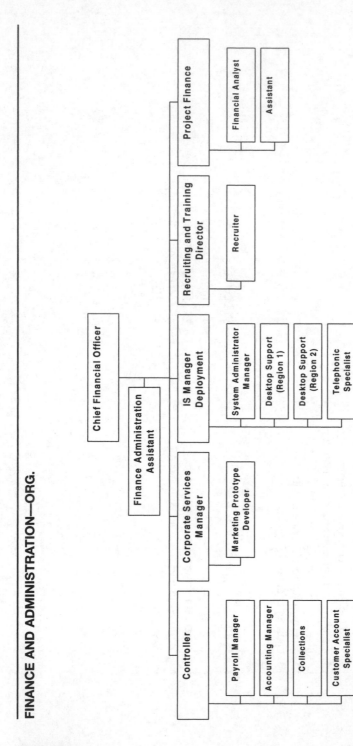

This is the final head count slide and the last slide in the presentation that the managers give. When they leave, Greg has private time to discuss issues with the board. With only two slides per department, RightNow's approach makes for a pretty short and to-the-point meeting. These slides can be viewed on the Entrepreneur America Web site (www.entrepreneur-america.com).

Who Is in the Room	Topic	Who Talks	Time
Board only	Administrative matters that need board approval, signatures, stock approvals	CEO	5 min. total
Board & Mgrs.	Each dept. reports on goals	Mgrs.	15 min. each
Board only	Closing: board talks bluntly with CEO	Board	15 min. total

Let me tell you what *not* to talk about at board meetings: strategic planning, marketing surveys, and product market positioning where you list all alternatives from A to Z. These discussions are certainly necessary, but not in real time in front of the entire board. Have them with your team beforehand, preferably spending a day or two off-site setting and analyzing goals. The results (and even the thinking behind setting the goals) are what to talk about with the board.

The main reason not to bring up this kind of open-ended stuff in meetings is that you'll get off-the-cuff reactions from board members. What you really want are studied, carefully considered responses. If you call them in advance, individually, each board member can spend time thinking seriously about the issues.

> **TIP**
>
> **The sales figures typically lead off board meetings. But at Ascend I always led with the department that was really struggling, the one that I wanted the board to focus on while they were still fresh and eager to focus. For us it was usually the quality service team. Last came departments that were fairly solid and under control—engineering and operations, in our case.**

Dealing with Change

Companies advised by sophisticated boards and hiring hotshot management teams aren't really start-ups anymore. Congratulations! You've made it to the big time. But along with the growth (and growing pains) often comes a little of what I call "postfunding blues." This syndrome usually attacks founders after the money arrives and the new professional management recruits join the team. The symptoms include whining that "things have changed."

Well, yes, things have changed. You've gotten money, management, and your business off the ground. A certain amount of change comes with the transition from scrappy start-up to promising growth company.

Here are some typical funks that funded start-ups fall into and what to do about them.

PRODUCT LOVE

It's a natural instinct to fall madly in love with your product or service. But listen to what customers and investors say about it, and be honest with yourself. Maybe it's not the right one for you.

Symptoms: When you talk about the concept, people are confused or aren't engaged. Or the product is selling but not even close to spiking. It's holding, at best. Customers repeatedly say that you're not a full solution for them, just a piece.

Cure: Pull back and regroup. Study your Sunflower Model and think about what direction to pursue next. Talk to customers about your ideas, and work their feedback into your idea before pursuing it full-bore.

WE MISS THE CEO

This syndrome strikes when the successful CEO starts hitting the road, visiting vendors, customers, and investors. The CEO office sits empty most of the time, instead of being open for casual visits as in the start-up phase.

Symptoms: Employees start to miss the camaraderie of the "old days," complaining that they never get to see the CEO anymore.

Cure: Senior people can spend a week tagging along on the road trip. Not only will employees grab some valuable face time, but they can learn a lot from sitting in while the CEO meets with investors, customers, and others. They'll probably end up with "hotel shock" and realize how heroic the CEO is for putting up with it. Greg took his co-founder out on trips.

For everyone else: Get over it! Once people are all piled up with work they tend to stop complaining about missing the CEO. This happened at Virtual Ink—everyone missed hanging out with the founder. Then the company got really successful and everyone was suddenly too busy to complain.

MY BABY IS ALL GROWN UP!

This syndrome strikes after the first product or service closes and ships. This first "baby," the labor of love for all employees, has finally gone out in the world.

Symptoms: "Planning attacks"—massive anxiety attacks about what to do next; numerous planning meetings.

Cure: Have a big send-off party. Then focus on making it work for customers. Go out, spend time with customers, and listen to their feedback. Fix any problems quickly, and spin out new features.

Most start-ups *avoid* the pain of change as long as they can. This always proves fatal because companies stick with dead-end products and strategies. Waiting around in the face of change (instead of taking decisive action) also proves fatal because the company often drifts along with no focus. Employees give up and quit. Investors lose confidence. Customers start to perceive your company as a lost cause instead of a solid vendor.

By now your start-up is getting to be pretty grown-up. Maybe the fat paycheck isn't quite there yet, but you've got a little growth capital. If you manage to leverage it well, before long you'll be the one driving the red Ferrari to work (you should probably give one of your board members a ride). Don't forget the vanity license plates.

EXERCISE:

1. As the CEO, prepare a list of major initiatives for your company. Break them into the following management disciplines, listing three major initiatives per area. The goals should cover three months of operations:

a) engineering

b) marketing

c) sales

d) customer support

e) operations

f) business development

2. Create an organizational chart for your company. Use dotted lines for unfilled positions.

3. For each department, make a list of the top two key people you need to hire. Include the date by which you need them and one or two sentences describing what characteristics you are looking for.

▶ You are welcome to refer to the Entrepreneur America Web site for additional information (www.entrepreneur-america.com).

Epilogue:
Companies That Make It

W hen a company comes out to Entrepreneur America and really impresses me, I get involved in helping it through the chaotic early stages. Typically the start-ups I see ask me to sit on their boards. But between running my ranch and the Entrepreneur America program, I just don't have time to sign up with everyone. I only sit on the boards of the ones I think I can really help. Currently that's fifteen companies.

As you've probably learned from reading the book, what I do with the start-ups is not so different from what venture capitalists do (except for investing in them). I ask probing questions, force them to focus, find the holes in their plans, and push them to the next level. How well and quickly they respond puts my Wanna-be teams into one of two categories.

HOT AND GETTING HOTTER

For companies in this category, things are humming. They've mastered dancing smoothly through the chaos. The business model has been proven by testing it on customers. The company earns several million in revenue or is poised for a major alliance with rapid growth expected. The executive team is in place, and the product or service is considered to be an industry leader. In short, the company either has or shortly will have an initial public offering. I've got six start-ups in this position.

LookSmart went public in early 2000 with Goldman, Sachs as the lead banker. By spring the company (listed on Nasdaq as LOOK) was trading at around $64 a share with a market cap of $8 billion. It has revenues of around $100 million.

The defining moment for this company was the reinvention of the business model. LookSmart originally intended to sell content to consumers as a Web portal company (competing with Yahoo! and others). This was their biggest problem—how to stand out from the crowd. It wasn't clear how the company would make money (the advertising model is too unstable) or scale.

As we worked on their Sunflower Model, we decided instead to focus on licensing and distributing content to businesses. After that the company quickly landed Microsoft as a client. That was the "IPO moment." LookSmart went on to make many other alliances with major companies, including British Telecom and Time Warner.

As I write this in early 2000, RightNow Technologies is slated to go public in mid-2000 with Credit Suisse First Boston. RightNow, which makes customer-relation software and services e-commerce Web sites, has revenues in excess of $5 million and is growing rapidly with projections of $20 million plus.

RightNow's biggest challenge was location, location, location. Based in Bozeman, Montana, the company had to work hard to recruit top-level management (mostly convincing spouses that Montana is a wonderful place to live). Realizing that Montana isn't for everyone, RightNow has begun to open distribution offices around the world. The company wasn't on VC radar, which actually turned out to be a good thing. CEO Greg Gianforte was forced to grow the company on cash flow.

I can't say enough good things about this company.

Greg's Guts and Brains team did everything right, buckling down and working hard to establish the company as one of its industry leaders. The reason this company will make it is that they've helped define an entirely new market. With their software product they've hit the ball solidly out of the park.

Virtual Ink created a new category with their Mimio product (the hardware/software combination that turns whiteboards into PC input devices). Great product, but the real key to the company's success is that CEO Greg McHale is excellent at closing large deals. He basically locked up his industry's distribution channels because he was able to sign exclusive contracts with large retail clients.

His biggest problem was in getting those channels moving. Once Greg was signed on with his large distributors, he had to work hard to prime the pump and build demand for the Mimio product. Also, he had to concentrate on being more than a one-product company. Greg has worked his Sunflower Model hard to churn out a slew of ideas for the coming year. His competition is getting tough, and that forces Virtual Ink to be better.

At the end of 2000 Creditland is expected to go public. The company, which established an on-line "financial superstore" that helps make credit decisions, has won some very significant alliances with the likes of Bank of America, MBNA, and others. The idea is ambitious. Creditland is approaching it by developing key partnerships and excellent core technology of the credit decision engine.

Founder Tony Wilbert is one of those rare finds—a natural entrepreneur. He's extremely skilled at recognizing quality people and recruiting them. Tony is brilliant at attracting a lot of young superstars. But none of them have experience in growing a company. It still remains to be seen if the team can pull that off. Creditland has also yet to prove that the business model can earn serious revenues and profits.

A little further behind the others, Virtmed is nonetheless off to a good start. The company has just brought on board an all-star management team. It's still not clear if these big-company recruits are entrepreneurial enough to launch a start-up, or if they can work together as a team, but I'm confident they can pull it off.

It took the team a year to secure funding after they first visited me at Entrepreneur America, but they never gave up. Once capitalized by New Enterprise Associates and Frazier & Company, they used the money to build the powerful management team and test-market and fine-tune their PalmPilot-based billing system with a few prominent health organizations. They are on the verge of striking several large deals. If they land those clients, they're on track for an IPO.

MOVING RIGHT ALONG

This category is where the bulk of my start-ups reside. Sometimes companies visit me, hoping that I'll make magic happen and instantly set them up for an IPO. Instead I usually send them away with lots of homework— months and months of homework sometimes. The companies in the "hot and getting hotter" category succeeded because they did that homework and did it well.

The companies in this "moving right along" category are still in homework mode. In general they have (or soon will have) a product, they may earn some revenues, the business model is beginning to emerge as an easily explainable model, and the management team is still somewhat incomplete. These companies are on the verge of their first venture capital financing. Generally they already have some angel money.

Silicon Spice is closest to jumping into the "hot" category. The semiconductor company is producing a new family of chips for telecommunications markets. It has a powerful architecture and software with over 150 man-years of effort—a serious barrier to entry.

As mentioned in chapter 7, after a year the start-up landed Vinod Dham, the father of the Intel processor. Once Dham came on board, the company won $30 million in funding, which grew to a total of $70 million. That's a real coup, but there are still some challenges ahead. The biggest is that they're building a very complicated product, and it's aimed at only a handful of customers. It's very difficult to nudge your way into supplying a company like Cisco. Silicon Spice's future might be in a merger or acquisition. I'm confident that the "class A" team will put the company in good shape for whatever future lies ahead.

Passlogix is the next closest to moving up. The company creates software that enables businesses to manage and verify the electronic identities of their customers and employees. The company has diligently marketed its story to name-brand accounts throughout Wall Street, but the problem is that it still hasn't condensed the pitch into an articulate one-page business plan.

Passlogix received several million dollars of seed angel money and is poised for its first venture capital round. But to land that, the company will have to develop a single clear statement that sums up what it does. At some point I'm sure they'll find that New York City is an expensive place to grow an engineering team. Everything is fine so far, but the company is still small.

Vellis produces software and services for e-commerce and training in the automotive industry. The company has finished the hard work of testing and honing the product with clients. Several prestigious customers have signed on, giving Vellis real weight with the VC world. The business model is strong and virile and offers a powerful value proposition. Vellis has gotten about $6 million in capital.

It's still not clear whether or not the founder will make a strong CEO. This company is his vision, and he left Australia for Chicago to see it through; but he doesn't have any CEO experience. He's very good with customers, and the board is hoping that he'll have time to learn other skills before the company really needs them.

Some of my early-stage start-ups have recently secured their first funding rounds. Actuality builds three-dimensional displays for use with CAD software. The company's product pushes the edge of the technology and will probably create a whole new market. This past spring Actuality secured $1.5 million in seed funding. But the company is still very much in the throes of discovering whether or not the dogs like the dog food (chapter 2). I love the founders, but the idea is still very iffy in the business sense.

Netcracker recently received seed financing of $500,000. The company is building a Web portal to facilitate inventory management of networking hardware equipment. They have spent years building a proprietary database that lists network parts—a significant barrier to entry.

The company is at a crossroads right now. Customer research is pointing strongly toward a particular market and application. Netcracker has to decide whether to pursue that (and spend the next three months attempting to raise money) or spend a little more time in customer research before making the commitment.

If this market is as strong as the initial research indicates, the company might be able to grow simply on cash flow. That would be better for the founder, who wants to run the company but would probably get replaced by investors looking for more savvy leadership. The founder has to decide if stepping aside would grow his company faster than remaining at the helm.

Everfile is a start-up with a very experienced engineering team, working to build an extranet-type service for small and medium-size companies. They would be able to share files and documents over the Web. The team has worked diligently on the business model and finally closed on $1.2 million in angel funding. The goal now is to get some strong sales and marketing expertise into the company. The CEO just doesn't handle customers well in a sales situation.

You can keep up with these companies (and new Wanna-bes I take on) by checking in at the Entrepreneur America's Web site at www.entrepreneur-america.com. You'll notice that there is one key way in which my work at Entrepreneur America differs from what investors do: I put energy into companies at earlier stages. A venture capitalist won't spend much time on a young start-up like Everfile or AIM. Once they decide not to invest, VCs just walk away. They're too busy to coach young teams to victory. I admit that most of my energy goes into the pre-IPO companies. The ones that are closest to succeeding deserve the most attention—after all, they've proven to me that they can do the hard work required to reach the finish line.

But if a young company shows me they're strong and ready to grow, I'll give them some fertilizer. They have to do something productive with it, not just toss it out there and hope it feeds them. When I see a company—no matter how small or unsophisticated—take my advice and work hard, focused on growing strong, then they get my attention. One thing I've learned after working closely with dozens of successful start-ups is that if you have guts and brains, you're going to make it.

Index